SO-BZF-418

London Mathematical Society Student Texts

Managing editor: Professor J. W. Bruce, Department of Mathematics
University of Liverpool, United Kingdom

Logic, Induction and Sets

THOMAS FORSTER
University of Cambridge

PUBLISHED BY THE PRESS SYNDICATE OF THE UNIVERSITY OF CAMBRIDGE
The Pitt Building, Trumpington Street, Cambridge, United Kingdom

CAMBRIDGE UNIVERSITY PRESS
The Edinburgh Building, Cambridge CB2 2RU, UK
40 West 20th Street, New York, NY 10011-4211, USA
477 Williamstown Road, Port Melbourne, VIC 3207, Australia
Ruiz de Alarcón 13, 28014 Madrid, Spain
Dock House, The Waterfront, Cape Town 8001, South Africa

http://www.cambridge.org

First published 2003

Printed in the United States of America

Typeface Computer Modern   *System* LATEX   [AU]

A catalog record for this book is available from the British Library.

Library of Congress Cataloging in Publication Data

Forster, Thomas, 1948–
    Logic, induction and sets / Thomas Forster.
        p. cm. – (London Mathematical Society student texts ; 56)
    Includes bibliographical references and index.
    ISBN 0-521-82621-7 – ISBN 0-521-53361-9 (pb.)
    1. Axiomatic set theory. I. Title. II. Series.

QA248 .F69 2003
511.3'22—dc21   2002041001

ISBN 0 521 82621 7 hardback
ISBN 0 521 53361 9 paperback

Contents

Preface

When there are so many textbooks on logic already available, an author of a new one must expect to be challenged for explanations as to why he has added to their number. I have four main excuses. I am not happy with the treatments of well-foundedness nor of the axiomatisation of set theory in any of the standard texts known to me. My third excuse is that, because my first degree was not in mathematics but in philosophy and music, I have always been more preoccupied with philosophical concerns than have most of my colleagues. Both the intension-extension distinction and the use-mention distinction are not only philosophically important but pedagogically important too: this is no coincidence. Many topics in mathematics become much more accessible to students if approached in a philosophically sensitive way. My fourth excuse is that nobody has yet written an introductory book on logic that fully exploits the expository possibilities of the idea of an inductively defined set or recursive datatype. I think my determination to write such a book is one of the *sequelæ* of reading Conway's beautiful book (2001) based on lectures he gave in Cambridge many years ago when I was a Ph.D. student.

This book is based on my lecture notes and supervision (tutorial) notes for the course entitled "Logic, Computation and Set Theory", which is lectured in part II (third year) of the Cambridge Mathematics Tripos. The choice of material is not mine, but is laid down by the Mathematics Faculty Board having regard to what the students have learned in their first two years. Third-year mathematics students at Cambridge have learned a great deal of mathematics, as Cambridge is one of the few schools where it is possible for an undergraduate to do nothing but mathematics for three years; however, they have done no logic to speak of. Readers who know more logic and less mathematics than did the original audience for this material – and among mathematicians they may well be a majority outside these islands – may find the emphasis rather odd. The part IIb course, of which this is a component, is designed for strong mathematics students who wish to go further and who need some exposure to logic: it was never designed to produce logicians. This book was written to meet a specific need, and to those with that need I offer it in the hope that it can be of help. I offer it also in the hope that it will convey to mathematicians something of the flavour of the distinctive way logicians do mathematics.

Like all teachers, I owe a debt to my students. Any researcher needs students for the stimulating questions they ask, and those attempting to write textbooks will be grateful to their students for the way they push us to give clearer explanations than our unreflecting familiarity with elementary material normally generates. At times students' questions will provoke us into saying things we had not realised we knew. I am also grateful to my colleagues Peter Johnstone and Martin Hyland for exercises they provided.

Introduction

In the beginning was the Word, and the Word was with God, and the Word
was God. The same was also in the beginning with God.

John's Gospel, ch 1 v 1

Despite having this text by heart I still have no idea what it means.
What I do know is that the word that is translated from the Greek into
English as 'word' is $\lambda o \gamma o \sigma$, which also gave us the word 'logic'. It is
entirely appropriate that we use a Greek word since it was the Greeks
who invented logic. They also invented the axiomatic method, in which
one makes basic assumptions about a topic from which one then derives
conclusions.

The most striking aspect of the development of mathematics in its
explosive modern phase of the last 120-odd years has been the extension of the scope of the subject matter. By this I do not mean that
mathematics has been extended to new subject areas (one thinks immediately of the way in which the social sciences have been revolutionised
by the discovery that the things they study can be given numerical values), even though it has, nor do I mean that new kinds of mathematical
entities have been discovered (imaginary numbers, vectors and so on),
true though that is too. What I mean is that in that period there was a
great increase in the variety of mathematical entities that were believed
to have an independent existence.

To any of the eighteenth-century mathematicians one could have begun an exposition "Let n be an integer..." or "Let n be a real..." and
they would have listened attentively, expecting to understand what was
to come. If, instead, one had begun "Let f be a set of reals ..." they
would not. The eighteenth century had the idea of an *arbitrary integer*
or an *arbitrary point* or an *arbitrary line*, but it did not have the idea of
an *arbitrary real valued function*, or an *arbitrary set of reals*, or an ar-

1

bitrary set of points. During this period mathematics acquired not only the concept of an arbitrary real-valued function, but also the concepts of arbitrary set, arbitrary formula, arbitrary proof, arbitrary computation, and additionally other concepts that will not concern us here. A reader who is not happy to see a discussion begin "Let x be an arbitrary ...", where the dots are to be filled in with the name of a suite of entities (reals, integers, sets), is to a certain extent not admitting entities from that suite as being fully real in the way they admit entities whose name they will accept in place of the dots. This was put pithily by Quine: "To be is to be the value of a variable". There are arbitrary X's once you have made X's into mathematical objects.

At the start of the third millenium of the common era, mathematics still has not furnished us with the idea of an arbitrary game or arbitrary proof. However, there is a subtle difference between this shortcoming and the eighteenth century's lack of the concept of an arbitrary function. Modern logicians recognise the lack of a satisfactory formalisation of a proof or game as a shortcoming in a way in which the eighteenth century did not recognise their lack of a concept of arbitrary function.

This historical development has pedagogical significance, since most of us acquire our toolkit of mathematical concepts in roughly the same order that the western mathematical tradition did. Ontogeny recapitulates phylogeny after all, and many students find that the propensity to reason in a freewheeling way about arbitrary reals or functions or sets does not come naturally. The ontological toolkit of school mathematics is to a large extent that of the eighteenth century. I remember when studying for my A-level being nonplussed by Richard Watts-Tobin's attempt to interest me in Rolle's theorem and the intermediate value theorem. It was too general. At that stage I was interested only in specific functions with stories to them: $\Sigma_{n \in \mathbb{N}} x^{2^n}$ was one that intruiged me, as did the function $\Sigma_{n \in \mathbb{N}} x^n \cdot n!$ in Hardy's (1949), which I encountered at about that time. I did not have the idea of an arbitrary real-valued function, and so I was not interested in general theorems about them.

Although understanding cannot be commanded, it will often come forward (albeit shyly) once it becomes clear what the task is. The student who does not know how to start answering "How many subsets does a set with n elements have?" may perhaps be helped by pointing out that their difficulty is that they are less happy with the idea of an arbitrary set than with the idea of an arbitrary number. It becomes easier to make the leap of faith once one knows which leap is required.

Some of these new suites of entities were brewed in response to a need to solve certain problems, and the suites that concern us most will be those that arose in response to problems in logic. Logic exploded into life in the twentieth century with the Hilbert programme and the celebrated incompleteness theorem of Gödel. It is probably a gross simplification to connect the explosive growth in logic in the twentieth century with the Hilbert programme, but that is the way the story is always told. In his famous 1900 address Hilbert posed various challenges whose solution would perforce mean formalising more mathematics. One particularly pertinent example concerns Diophantine equations, which are equations like $x^3 + y^5 = z^2 + w^3$, where the variables range over integers. Is there a general method for finding out when such equations have solutions in the integers? If there is, of course, one exhibits it and the matter is settled. If there is not, then in order to prove this fact one has to be able to say something like: "Let $\mathcal{A}$ be an arbitrary algorithm ..." and then establish that $\mathcal{A}$ did not perform as intended. However, to do *that* one needs a concept of an algorithm as an arbitrary mathematical object, and this was not available in 1900. It turns out that there is no method of the kind that Hilbert wanted for analysing diophantine equations, and in chapter 6 we will see a formal concept of algorithm of the kind needed to demonstrate this.

This extension of mathematical notation to nonmathematical areas has not always been welcomed by mathematicians, some of whom appear to regard logic as mere notation: "If Logic is the source of a mathematician's hygiene, it is not the source of his food" is a famous sniffy aside of Bourbaki. Well, one *bon mot* deserves another: there is a remark of McCarthy's as famous among logicians as Bourbaki's is to mathematicians to the effect that, "It is reasonable to hope that the relationship between computation and mathematical logic will be as fruitful in the next century as that between analysis and physics in the last." With this at the back our minds it has to be expected that when logicians write books about logic for mathematicians they will emphasise the possible connections with topics in theoretical computer science.

The autonomy of syntax

One of the great insights of twentieth-century logic was that, in order to understand how formulæ can bear the meanings they bear, we must first strip them of all those meanings so we can see the symbols as themselves. Stripping symbols of all the meanings we have so lovingly bestowed on

them over the centuries in various unsystematic ways[1] seems an extremely perverse thing to do – after all, it was only so that they could
bear meaning that we invented the symbols in the first place. But we
have to do it so that we can think about formulæ as (perhaps mathematical) objects in their own right, for then can we start to think about
how it is possible to ascribe meanings to them in a systematic way that
takes account of their internal structure. That makes it possible to prove
theorems about what sort of meanings can be born by languages built
out of those symbols. These theorems tend to be called *completeness
theorems*, and it is only a slight exaggeration to say that logic in the
middle of the twentieth century was dominated by the production of
them.

It is hard to say what logic is dominated by now because no age
understands itself (a very twentieth century insight!), but it does not
much matter here because all the material in this book is fairly old
and long-established. All the theorems in this will be older than the
undergraduate reader; most of them are older than the author.

Finally, a cultural difference. Logicians tend to be much more concerned than other mathematicians about the way in which desirable
propositions are proved. For most mathematicians, most of the time,
it is enough that a question should be answered. Logicians are much
more likely to be concerned to have proofs that use particular methods,
or refrain from exploiting particular methods. This is at least in part
because the connections between logic and computation make logicians
prefer proofs that correspond to constructions in a way which we will see
sketched later, but the reasons go back earlier than that. Logicians are
more likely than other mathematicians to emphasise that 'trivial' does
not mean 'unimportant'. There are important trivialities, many of them
in this book. The fact that something is unimportant may nevertheless
itself be important. There are some theorems that it is not a kindness
to the student to make seem easy. Some hard things should be seen to
be hard.

---

[1] The reader is encourged to dip into Cajori's *History of Mathematical Notations*
to see how unsystematic these ways can be and how many dead ends there have
been.

1

Definitions and notations

This chapter is designed to be read in sequence, not merely referred back to. There are even exercises in it to encourage the reader.

Things in **boldface** are usually being **defined**. Things in *italic* are being *emphasised*. Some exercises will be collected at the end of each chapter, but there are many exercises to be found in the body of the text. The intention is that they will all have been inserted at the precise stage in the exposition when they become doable.

I shall use lambda notation for functions. $\lambda x.F(x)$ is the function that, when given x, returns $F(x)$. Thus $\lambda x.x^2$ applied to 2 evaluates to 4. I shall also adhere to the universal practice of writing '$\lambda xy.(\ldots)$' for '$\lambda x.(\lambda y.(\ldots))$'. Granted, most people would write things like '$y = F(x)$' and '$y = x^2$', relying on an implicit convention that, where 'x' and 'y' are the only two variables are used, then y is the output ("ordinate") and x is the input ("abcissa"). This convention, and others like it, have served us quite well, but in the information technology age, when one increasingly wants machines to do a lot of the formula manipulations that used to be done by humans, it turns out that lambda notation and notations related to it are more useful. As it happens, there will not be much use of lambda notation in this text, and I mention it at this stage to make a cultural point as much as anything. By the same token, a word is in order at this point on the kind of horror inspired in logicians by passages like this one, picked almost at random from the literature (Ahlfors, 1953 p. 69):

Suppose that an arc γ with equation $z = z(t), \alpha \leq t \leq \beta$ is contained in a region Ω, and let f be defined and continuous in Ω. Then $w = w(t) = f(z(t))$ defines an arc ...

The linguistic conventions being exploited here can be easily followed

by people brought up in them, but they defy explanation in any terms
that would make this syntax machine-readable. Lambda notation is
more logical. Writing '$w = \lambda t.f(z(t))$' would have been much better
practice.

I write ordered pairs, triples, and so on with angle brackets: $\langle x, y \rangle$. If
x is an ordered pair, then $\mathtt{fst}(x)$ and $\mathtt{snd}(x)$ are the first and second
components of x. We will also write '$\vec{x}$' for '$x_1 \ldots x_n$'.

1.1 Structures

A set with a relation (or bundle of relations) associated with it is called
a **structure**, and we use angle brackets for this too. $\langle X, R \rangle$ is the set X
associated with the relation R, and $\langle X, R_1, R_2 \ldots R_n \rangle$ is X associated
with the bundle of relations – $R_1 \ldots R_n$. For example, $\langle \mathbb{N}, \leq \rangle$ is the
naturals as an ordered set.

The elements are "in" the structure in the sense that they are members
of the underlying set – which the predicates are not. Often we will use
the same letter in different fonts to denote the structure and the *domain*
of the structure; thus, in "$\mathfrak{M} = \langle M, \ldots \rangle$" M is the domain of $\mathfrak{M}$. Some
writers prefer the longer but more evocative locution that M is the
carrier set of $\mathfrak{M}$, and I will follow that usage here, reserving the word
'**domain**' for the set of things that appear as elements of n-tuples in R,
where R is an n-place relation. We write '$dom(R)$' for short.

Many people are initially puzzled by notations like $\langle \mathbb{N}, \leq \rangle$. Why specify the ordering when it can be inferred from the underlying set? The
ordering of the naturals arises from the naturals in a – natural(!) – way.
But it is common and natural to have distinct structures with the same
carrier set. The rationals-as-an-ordered-set, the rationals-as-a-field and
the rationals-as-an-ordered-field are three distinct structures with the
same carrier set. Even if you are happy with the idea of this distinction
between carrier-set and structure and will not need for the moment the
model-theoretic jargon I am about to introduce in the rest of this paragraph, you may find that it helps to settle your thoughts. The rationals-
as-an-ordered-set and the rationals-as-an-ordered-field have the same
carrier set, but different signatures (see page 48). We say that the
rationals-as-an-ordered-field are an **expansion** of the rationals-as-an-
ordered-set, which in turn is a **reduction** of the rationals-as-an-ordered-
field. The reals-as-an-ordered-set are an **extension** of the rationals-
as-an-ordered-set, and, conversely, the rationals-as-an-ordered-set are a
substructure of the reals. Thus:

Beef up the signature to get an *expansion.*
Beef up the carrier set to get an *extension.*
Throw away some structure to get a *reduction.*
Throw away some of the carrier set to get a *substructure.*

We will need the notion of an **isomorphism** between two structures.
If $\langle X, R \rangle$ and $\langle Y, S \rangle$ are two structures, they are **isomorphic** iff there
is a bijection f between X and Y such that, for all $x, y \in X$, $R(x, y)$ iff
$S(f(x), f(y))$.

(This dual use of angle brackets for tupling and for notating structures
has just provided us with our first example of **overloading**. "Overloading"!? It is computer science-speak for "using one piece of syntax
for two distinct purposes" – commonly and gleefully called "abuse of
notation" by mathematicians.)

1.2 Intension and extension

Sadly the word 'extension', too, will be overloaded. We will not only
have extensions of models – as just now – but extensions of theories
(of which more later), and there is even **extensionality**, a property of
relations. A binary relation R is extensional if $(\forall x)(\forall y)(x = y \longleftrightarrow
(\forall z)(R(x, z) \longleftrightarrow R(y, z))$. Notice that a relation can be extensional
without its converse (converses are defined on page 9) being extensional:
think "square roots". An extensional relation on a set X corresponds
to an injection from X into $\mathcal{P}(X)$, the power set of X. For us the most
important example of an extensional relation will be $\in$, set membership.
Two sets with the same members are the same set.

Finally, there is the intension extension distinction, an informal device but a standard one we will need at several places. We speak of
functions-in-intension and **functions-in-extension** and in general
of **relations-in-intension** and **relations-in-extension**. There are also
'intensions' and 'extensions' as nouns in their own right.

The standard illustration in the literature concerns the two properties
of being *human* and being a *featherless biped* – a creature with two legs
and no feathers. There is a perfectly good sense in which these concepts
are the same (one can tell that this illustration dates from before the
time when the West had encountered Australia with its kangaroos!), but
there is another perfectly good sense in which they are different. We
name these two senses by saying that 'human' and 'featherless biped'
are the same property in extension but different properties in intension.

A more modern and more topical illustration is as follows. A piece of code that needs to call another function can do it in either of two ways. If the function being called is going to be called often, on a restricted range of arguments, and is hard to compute, then the obvious thing to do is compute the set of values in advance and store them in a look-up table in line in the code. On the other hand if the function to be called is not going to be called very often, and the set of arguments on which it is to be called cannot be determined in advance, and if there is an easy algorithm available to compute it, then the obvious strategy is to write code for that algorithm and call it when needed. In the first case the embedded subordinate function is represented as a function-in-extension, and in the second case as a function-in-intension. Functions-in-extension are sometimes called the **graph**s of the corresponding functions-in-intension: the graph of a function f is $\{\langle x, y \rangle : x = f(y)\}$. One cannot begin to answer exercise 1(vi) unless one realises that the question must be, "How many binary relations-*in-extension* are there on a set with n elements?" (There is no answer to "how many binary relations-in-intension ... ")

I remember being disquieted – when I was a A-level student – by being shown a proof that if one integrates $\lambda x.\frac{1}{x}$ with respect to x, one gets $\lambda x.log(x)$. The proof procedes by showing that the two functions are the same function-in-extension – or at least that they are both roots of the one functional equation, and that did not satisfy me.

The intension – extension distinction is not a formal technical device, and it does not need to be conceived or used rigorously, but as a piece of mathematical slang it is very useful. One reason why it is a bit slangy is captured by an *aperçu* of Quine's: "No entity without identity". What this *obiter dictum* means is that if you wish to believe in the existence of a suite of entities – numbers, ghosts, functions-in-intension or whatever it may be – then you must have to hand a criterion that tells you when two numbers (ghosts, functions-in-intension) are the same number (ghost, etc.) and when they are different numbers (ghosts, etc). We need *identity criteria* for entities belonging to a suite before those entities can be used rigorously. And sadly, although we have a very robust criterion of identity for functions-in-extension, we do not yet have a good criterion of identity for functions-in-intension. Are the functions-in-intension $\lambda x.x + x$ and $\lambda x.2 \cdot x$ two functions or one? Is a function-in-intension an algorithm? Or are algorithms even more intensional than functions-in-intension?

Finally, this slang turns up nowadays in the connection with the idea of evaluation. In recent times there has been increasingly the idea that intensions are the sort of things one *evaluates* and that the things to which they evaluate are extensions.

1.3 Notation for sets and relations

Relations in extension can be thought of as sets of ordered tuples, so we had better ensure we have to hand the elementary set-theoretic gadgetry needed.

Some people write '$\{x|F\}$' for the set of things that are F, but since I will be writing '$|x|$' for the cardinal of x, I shall stick to '$\{x : F(x)\}$'. This notation is commonly extended by moving some of the conditions expressed on the right of the colon to the left: for example, '$\{x \in \mathbb{N} : (\exists y)(x = 2 \cdot y)\}$' instead of '$\{x : x \in \mathbb{N} \wedge (\exists y)(x = 2 \cdot y)\}$'. There is a similar notation for the quantifiers: often one writes '$(\forall n \in \mathbb{N})(\ldots)$' instead of '$(\forall n)(n \in \mathbb{N} \rightarrow \ldots)$'. The reader is presumably familiar with '$\subseteq$' for subset of, but perhaps not with '$x \supseteq y$' (read 'x is a superset of y'): it means the same as $y \subseteq x$. $\mathcal{P}(x)$ is the **power set** of x: $\{y : y \subseteq x\}$. Set difference: $x \setminus y$ is the set of things that are in x but not in y. The symmetric difference: $x \Delta y$, of x and y, is the set of things in one or the other but not both: $(x \setminus y) \cup (y \setminus x)$. (This is sometimes written 'XOR', but we will reserve XOR for the corresponding propositional connective). Sumset: $\bigcup x := \{y : (\exists z)(y \in z \wedge z \in x)\}$; and intersection $\bigcap x := \{y : (\forall z)(z \in x \rightarrow y \in z)\}$. These will also be written in indexed form at times: $\bigcup_{i \in I} A_i$. The **arity** of a function or a relation is the number of arguments it is supposed to have. It is a significant but generally unremarked fact that one can do most of mathematics without ever having to consider relations of arity greater than 2. These relations are **binary**. The **composition** of two binary relations R and S, which is $\{\langle x, z \rangle : (\exists y)(\langle x, y \rangle \in R \wedge \langle y, z \rangle \in S)\}$, is notated '$R \circ S$'. $R \circ S$ is not in general the same as $S \circ R$: the sibling of your parent is probably not the parent of your sibling. (Miniexercise: how is it legally possible for them to be the same?)

$R \circ R$ is written R^2, and similarly R^n. The **inverse** or **converse** of R, written 'R^{-1}', is $\{\langle x, y \rangle : \langle y, x \rangle \in R\}$. However, do not be misled by this exponential notation into thinking that $R \circ R^{-1}$ is the identity. See exercise 1.

It is sometimes convenient to think of a binary relation as a matrix whose entries are **true** and **false**. This has an advantage, namely, that under this scheme the matrix product of the matrices for R and S is the matrix for $R \circ S$. (Take multiplication to be $\wedge$ and addition to be

∨). However, in principle this is not a good habit, because it forces one to decide on an ordering of the underlying set (rows and columns have to be put down in an order after all) and so is less general than the picture of binary relations-in-extension as sets of ordered pairs. It also assumes thereby that every set can be totally ordered, and this is a nontrivial consequence of the axiom of choice, a contentious assumption of which we will see more later. However, it does give a nice picture of converses: the inverse – converse of R corresponds to the transpose of the matrix corresponding to R, and the matrix corresponding to $R \circ S$ is the product of the two matrices in the obvious way.

A relation R is **transitive** if $\forall x \forall y \forall z \ R(x,y) \wedge R(y,z)Dz \rightarrow R(x,z)$ (or, in brief, $R^2 \subseteq R$). A relation R is **symmetrical** if $\forall x \forall y (R(x,y) \longleftrightarrow R(y,x))$ or $R = R^{-1}$. Beginners often assume that symmetrical relations must be reflexive . They are wrong, as witness "rhymes with", "conflicts with", "can see the whites of the eyes of", "is married to", "is the sibling of" and so on.

An **equivalence relation** is symmetrical, transitive and reflexive. An equivalence relation $\sim$ is a **congruence relation for** an n-ary function f if, whenever $x_i \sim y_i$ for $i \leq n$, then $f(\vec{x}) \sim f(\vec{y})$. (The notation "$\vec{x}$" abbreviates a list of variables, all of the shape 'x' with different subscripts.) A cuddly familiar example is integers mod k: congruence mod k is a congruence relation for addition and multiplication of natural numbers. (It is not a congruence relation for exponentiation: something that often confuses beginners.) We will need this again in sections 3.4 (on boolean algebra) and 5.7 (on ultraproducts) and in chapter 7, on transfinite arithmetic.

I have used the adjective 'reflexive' without defining it. A binary relation on a set X is **reflexive** if it relates every member of X to itself. (A relation is **irreflexive** if it is disjoint from the identity relation: note that irreflexive is not the same as not reflexive!) That is to say, R is reflexive iff $(\forall x \in X)(\langle x, x \rangle \in R)$. Notice that this means that reflexivity is not a property of a relation, but of the structure $\langle X, R \rangle$ of which the relation is a component.

This annoying feature of reflexivity (which irreflexivity does not share) is also exhibited by **surjectivity**, which is a property not of a function but a function-with-a-range. A function is surjective if every element of the range is a value. **Totality** likewise is a property of a function-and-an-intended-domain. A function f on a set X is total if it is defined for every argument in X.

Some mathematical cultures make this explicit, saying that a function is an ordered triple of domain, range and a set of ordered pairs. This notation has the advantage of clarity, but it has not yet won the day.

In contrast, injectivity of a function-in-extension is a property solely of the function-in-extension and not of the intended domain or range. A function is injective iff it never sends distinct arguments to the same value.

While on the subject of functions, a last notational point. In most mathematical usage the terminology '$f(x)$' is overloaded: it can denote either the value that the function f allocates to the argument x or the *set* of values that f gives to the arguments in the set x. Normally this overloading does not cause any confusion, because typically it is clear from the context which is meant. $f(\pi)$ is clearly a number and $f(\Re)$ a set of numbers. The give-away here is in the style of letter used for the argument. The human brain is very good at exploiting cues like this for useful information (as witness the convenience of the notation $\mathcal{M} = \langle M, R \rangle$, and the readability of the Ahlfors example on page 5) but there are circumstances in which contextual decoding doesn't work. If everything is a set (and in set theory – which we meet in chapter 8 – everything is, indeed, a set!), there is no way of telling which of the two $f(x)$ is: it could be either!

Accordingly we will use the following notation, which is nowadays standard: $f``x$ is the set of values $f$ allocates to the arguments in the set $x$, and $f(x)$ will continue to be the value that $f$ assigns to the argument $x$. Older books on set theory sometimes use the notation $f`x$ ("One apostrophe, one value"; $f``x$ is "*plural* apostrophe, *set* of values") for our $f(x)$, but this notation (due originally to Russell and Whitehead 1919) is now obsolescent. For us the commonest use of this notation in in settings like "$f``X \subseteq X$", which says that $x$ is **closed under** $f$. Of course we can talk about sets being closed under $n$-ary functions with $n > 1$, and if $g$ is an $n$-ary function ("a function of $n$ variables"), then $g``(X^n) \subseteq X$.

1.4 Order

Now for a number of ideas that emerge from the concept of *order*. Order relations obviously have to be transitive, and they cannot be symmetrical because then they would not distinguish things, would they? Indeed transitive relations that are symmetrical are called *equivalence* relations (as long as they are reflexive). So how do we capture this failure of symmetry? We start by noticing that, although an order relation must of course be transitive and cannot be symmetrical, it is not obvious whether we want it to be reflexive or want it to be irreflexive. Since orderings represent ways of *distinguishing* things, they do not have anything natural to say about whether things are related to themselves or not. Is x less than itself? Or not? Does it matter which way we jump? Reflection on your experience with $<$ and $\leq$ on the various kinds of numbers you've dealt with (naturals, integers, reals and rationals) will make you feel that it does not much matter. After all, in some sense $<$ and $\leq$ contain the same information about numbers (See exercise 1(xviii)). This intuition is sound, and in most of mathematics we can indeed go either way. (An exception is well-orderings, of which we will see more later.) These two ways give rise to two definitions.

(i) A **strict partial order** is irreflexive, transitive and asymmetrical. (A relation is **asymmetrical** if it cannot simultaneously relate x to y and y to x. This of course implies irreflexivity.)

(ii) A **partial order** is reflexive, transitive and ... well it cannot be asymmetrical because $x \leq x$. We need to weaken asymmetry to a condition that says that, if $x \neq y$, then not both $x \leq y$ and $y \leq x$. This condition, usually expressed as its contrapositive (see page 22) $(\forall xy)(x \leq y \wedge y \leq x \rightarrow x = y)$, is **antisymmetry** and is the third clause in the definition of partial order.

If R is a partial ordering of a set X, then $R \setminus \{\langle x, x \rangle : x \in X\}$ is a strict partial ordering of X, and if R is a strict partial ordering of a set X, then $R \cup \{\langle x, x \rangle : x \in X\}$ is a partial ordering of X. Thus each concept (partial order and strict partial order) can be defined in terms of the other. There is a scrap of logical slang that comes in handy here: we say that each can be defined if we take the other as **primitive**.

There is a temptation to infer from the fact that a speaker refers to a relation as a "partial" order that they must mean that the order is not total on the grounds that, if it were total, they would have said so. This **conversational implicature** is very useful in ordinary parlance but can mislead us here. When we describe a given binary relation as a 'partial order' (or 'strict partial order') we do not mean to preclude the possibility of the order in question being total. The word 'partial' is there (in the common name) because we wish to be able to call relations 'orders' even if for some x and y they fail to prefer x to y or y to x.

Total orders are special kinds of orders that never fail in this way. Again, they come in two flavours:

(i) A **strict total** order is a strict partial order that satisfies the extra condition $(\forall xy)(x < y \lor y < x \lor x = y)$. Because this condition says there are no more than three possibilities, it is called **trichotomy** (from two Greek words meaning *three* and to *cut* as in a-*tom*, lobo-*tomy*.)

(ii) A **total order** is a partial order with the extra condition $(\forall xy)(x \le y \lor y \le x)$. This property is called **connexity**, and relations bearing it are said to be **connected**. Overloading of this last word is a frequent source of confusion, so beware.

Thus trichotomy and connexity are related to each other the way antisymmetry and asymmetry are.

A **poset** $\langle X, \le_X \rangle$ is a set X with a partial ordering $<_X$. The expression "strict poset", which one might expect to see being used to denote a set-with-strict-partial-order, seems not to be used.

A **monotone** function from a poset $\langle A, \le_A \rangle$ to a poset $\langle B, \le_B \rangle$ is a function $f : A \to B$ such that $\forall xy(x \le_A y \to f(x) \le_B f(y))$.

EXERCISE 1 *In the following questions assume the carrier set is a fixed set X, let 1 be the identity relation restricted to X and let U be the universal (binary) relation on X, namely, $X \times X$.*

(i) *Show that*

    (a) $R \subseteq S \to R \circ T \subseteq S \circ T$;

    (b) $R \subseteq S \to T \circ R \subseteq T \circ S$;

    (c) $R \circ (S \cup T) = R \circ S \cup R \circ T$;

    (d) $(R \circ S)^{-1} = S^{-1} \circ R^{-1}$.

(ii)   (a) *Is $R \setminus R^{-1}$ antisymmetric?[1] Asymmetric?*

    (b) *Is $R \Delta R^{-1}$ symmetrical? Antisymmetric? Asymmetric?*

    (c) *Is the composition of two symmetrical relations symmetrical?*

(iii) *What can you say about a function from X to itself whose graph is (i) a reflexive relation? or (ii) a transitive relation? or (iii) a symmetrical relation?*

(iv) *Match up the properties in the left column with those in the right.*

$R^2 \subseteq R$;	*R is symmetrical*
$R \cap R^{-1} = \emptyset$;	*R is antisymmetric*
$R \cap R^{-1} = 1$;	*R is asymmetrical*
$R = R^{-1}$;	*R is a permutation*

---

[1] See definition below.

$$R \circ R^{-1} = 1; \qquad R \text{ is connected}$$
$$R \cup R^{-1} = U; \qquad R \text{ is transitive}$$

(v) *Which of the following are true?*
$$R \subseteq S \rightarrow R^{-1} \subseteq S^{-1};$$
$$R \subseteq S \rightarrow R^{-1} \supseteq S^{-1};$$
$$R = R^{-1} \rightarrow 1 \subseteq R.$$
Relational algebra in this style goes back to Russell and White-head (1919). Sophisticates can find a rather intruiging read in Smullyan (2001).

(vi) *How many binary relations are there on a set of size n?*

(vii) *How many of them are reflexive?*

(viii) *How many are fuzzies? (A* fuzzy *is a binary relation that is symmetric and reflexive.)*

(ix) *How many of them are symmetrical?*

(x) *How many of them are antisymmetric?*

(xi) *How many are total orders?*

(xii) *How many are trichotomous?*

(xiii) *How many are antisymmetric and trichotomous?*

(xiv) *There are the same number of antisymmetric relations as trichotomous. Prove this to be true without working out the precise number.*

(xv) *(For the thoughtful student) If you have done parts* (xiii) *and* (ix) *correctly, the answers will be the same. Is there a reason why they should be the same?*

(xvi) *Do not answer this question. How many partial orders are there on a set of size n?*

(xvii) *Do not answer this question. How many strict partial orders are there on a set of size n?*

(xviii) *Should the answers to the two previous questions be the same or different? Give reasons. (Compare this with your answer to question* (xv).)

(xix) *Show that the proportion of relations on a set with n members that are extensional tends to 1 as $n \rightarrow \infty$.*

The **restriction** of a relation R to a carrier set X (which is $R \cap X^n$, where n is the arity of R) is denoted by '$R|X$'. (N.B. Overloading of '$|$'.) A **chain** in a poset $\langle X, \leq_X \rangle$, is a total ordering $\langle X', \leq_X |X' \rangle$, where $X' \subseteq X$. In words: a chain in a poset is a subset totally ordered by the restriction of the order relation. An **antichain** in a poset is a subset of the carrier set such that the restriction of the order relation to it is the identity relation.

An **upper semilattice** is a poset $\langle X, \leq_X \rangle$ such that $(\forall x_1, x_2)(\exists y \geq_X x_1, x_2)(\forall z)(x_1 \leq_X z \wedge x_2 \leq_X z \rightarrow y \leq_X z)$. By antisymmetry, if there is such a y, it is unique, and we write it $x_1 \vee x_2$ and refer to it as the **supremum** (**sup** for short) or the **least upper bound** (**lub** for short, also known as **join**) of x_1 and x_2. A **lower semilattice** is a also a poset, $\langle X, \leq_X \rangle$, such that $(\forall x_1, x_2)(\exists y \leq_X x_1, x_2)(\forall z)(x_1 \geq_X z \wedge x_2 \geq_X z \rightarrow y \geq_X z)$. By antisymmetry, if there is such a y, it is unique, and we write it $x_1 \wedge x_2$ and refer to it as the **infimum** (or **inf** for short) or the **greatest lower bound** (or **glb** for short; also known as **meet**) of x_1 and x_2. A **lattice** is something that is both an upper and a lower semilattice. Some people apply the word 'lattice' only to things with a top and a bottom element, and we will adhere to this custom here. The thinking behind this decision is that, if one thinks of a lattice as a poset in which every finite set of elements has both a sup and an inf (which appears to follow by an easy induction from the definition given), then one expects the empty set to have a sup and an inf – it is finite after all. And manifestly the sup and inf of the empty set must be the bottom and the top element of the lattice (and yes, it is that way round, not the other: check it!) A **complete** upper (resp. lower) semilattice is an upper (resp. lower) semilattice $\langle X, \leq \rangle$, where *every* subset X' of X (not just finite ones) has a sup (resp. inf.). We write these sups and inf (or lubs and glbs) in the style $\bigvee X'$ (sup or lub) and $\bigwedge X'$ (inf or glb). Everyone agrees that a complete lattice must have a top element and a bottom element.

An easy induction shows that in a lattice every finite set of elements has a sup and an inf. Notice also that in any lattice the set of things above a given element is also a lattice. These things are sometimes called "upper sets".[1]

If $\langle X, \leq_X \rangle$ is a poset, a subset $X' \subseteq X$ of X is a **directed** subset if $(\forall x_1 x_2 \in X')(\exists x_3 \in X')(x_1 \leq_X x_3 \wedge x_2 \leq_X x_3)$. (So, for example, if $\langle X, \leq_X \rangle$ is a total order, every subset is directed). A **directed union** is the sumset of a directed set, and similarly directed sups.

$\langle X, \leq_X \rangle$ is a **complete partial order** if every subset has a sup. It follows immediately that every subset also has an inf, so a complete

---

[1] "In Spain, all the best upper sets do it, Lithuanians and Letts do it, let's do it, let's fall in love!"

poset is simply a complete lattice. $\langle X, \leq_X \rangle$ is a **chain-complete poset** if every chain has a sup.

A lattice is **distributive** if $\forall xyz(x \wedge (y \vee z) = ((x \wedge y) \vee (x \wedge z)))$. It is **dually distributive** if $\forall xyz(x \vee (y \wedge z) = ((x \vee y) \wedge (x \vee z)))$.

EXERCISE 2 *A* **partition** *of a set x is a family Π of pairwise disjoint nonempty subsets of x that collectively exhaust x. The nonempty subsets comprising the partition are called* **pieces**. *If Π_1 and Π_2 are partitions of x, we say that Π_1* **refines** *Π_2 if every piece of Π_1 is a subset of a piece of Π_2.*

Show that for any set X the collection of partitions of X is a complete lattice under refinement. Is it distributive?

If sups and infs always exist, we can introduce a notation for them: '$x \vee y$' for the sup and '$x \wedge y$' for the inf are both standard. '0' and '1' for the bottom and top element are standard but not universal: in some cultures the bottom element is written '$\perp$'. Using these notations we can write down the following axioms for lattices:

$(\forall xy)(x \vee (x \wedge y) = x)$;
$(\forall xy)(x \wedge (x \vee y) = x)$;
$(\forall xyz)(x \vee (y \vee z) = (x \vee y) \vee z)$;
$(\forall xyz)(x \wedge (y \wedge z) = (x \wedge y) \wedge z)$;
$(\forall xy)(x \vee y = y \vee x)$;
$(\forall xy)(x \wedge y = y \wedge x)$.

The axioms for 0 and 1 are:
$(\forall x)(x \vee 1 = 1)$;
$(\forall x)(x \wedge 1 = x)$;
$(\forall x)(x \wedge 0 = 0)$;
$(\forall x)(x \vee 0 = x)$.

None of these axioms mention the partial order! In fact, we can define $\leq$ in terms of $\wedge$ or $\vee$ by $x \leq y$ iff $(x \vee y) = y$ (or by $(x \wedge y) = x$). Readers should do the miniexercise of checking this for themselves.

A lattice is **complemented** if it has elements 1 ("top") and 0 (or $\perp$ "bottom") and a function (written in various ways) $\neg$ s.t. $\forall x((x \wedge \neg x = 0) \wedge (x \vee \neg x = 1))$. (Note overloading of '$\wedge$'!) A **boolean algebra** is a distributive complemented lattice.

1.5 Products

The product of two structures $\langle X, R \rangle$ and $\langle Y, S \rangle$ is the structure whose carrier set is $X \times Y$, with the binary relation defined "pointwise":

$$\langle X \times Y, \{\langle \langle t, u \rangle, \langle v, w \rangle \rangle : \langle t, v \rangle \in R \wedge \langle u, w \rangle \in S\}\rangle.$$

You have encountered this in products of groups, for example. You should make a note here (though we shall not make use of this until section 5.7) that we can form products of more than two things at a time, and we will write things like '$\prod_{i \in I} \mathcal{A}_i$' to mean a product of $\mathcal{A}$s indexed by a set I by analogy with our notation for indexed unions $\bigcup_{i \in I} A_i$ earlier. $X \times X$ will often be denoted by 'X^2' and products of yet more copies of X denoted by 'X^n'.

If $\langle A, \leq_A \rangle$ and $\langle B, \leq_B \rangle$ are two posets, we can define a partial order on functions from A to B by setting $f \leq g$ iff $(\forall a \in A)(f(a) \leq g(a))$. We write '$A \to B$' for the set of all functions from A to B. Overloading of '$\to$' in this way is no mere overloading: it is a divine ambiguity, known as the *Curry-Howard correspondence*, on which a wealth of ink has been spilled. Try, for example, Girard Lafont and Taylor (1989)

Now if $\langle X, \leq_X \rangle$ and $\langle Y, \leq_Y \rangle$ are two partial orders, then we can define partial orders on $X \times Y$ in several ways. The product defined above is called the **pointwise** product. In the **lexicographic** order of the product we set $\langle x, y \rangle \leq_{lex} \langle x', y' \rangle$ if $x <_X x'$ or $x = x'$ and $y \leq_Y y'$. Although straightforward examples of lexicographic products are scarce, there are a number of combinatorial devices that have the flavour of a lexicographic product. One example is the Olympic league table: one grades nations in the first instance by the number of gold medals their athletes have won, then by the number of silvers and only if these fail to discriminate between them does one count the number of bronzes. In this setting there are three preorders on the set of nations (they are preorders because antisymmetry cannot be guaranteed: two nations may have the same number of medals of any given colour), and we are combining these three preorders into one preorder on the set of nations.

Other examples include the devices used to determine which team goes forward from a qualifying group in world cup football. *Prima facie* this should be the team with the largest number of points, but if two teams have the same number of points, one looks at the number of goals the two teams have scored, and so on, examining the values the two teams take under a sequence of parameters of dwindling importance. In cricket the analysis of a bowler who takes x wickets while conceding y runs is preferred to that of a bowler who takes x' wickets while conceding y' runs as long as $x > x'$ or $x = x' \wedge y < y'$. However, in none of these naturally occurring cases is one ordering *tuples* of things: rather, one is trying to order things by combining in various ways various preorders of the things. However, the underlying intuition is the same.

Notice that the lexicographic product is a superset of the pointwise product. If we have two partial orders with the same carrier set and (the graph of, or extension of) one is a superset of (the graph of, or extension of) the other, we say the first **extends** the second. The **colex** ordering of $X \times Y$ orders pairs according to *last* difference. The colex ordering too is a superset of the pointwise product ordering. In fact, the pointwise product ordering is the intersection of the lexicographic ordering and the colex ordering. Miniexercise: Check this.

One naturally tends to think of preorders as preference orders, as the preorders in the illustrations above of course are. Although naturally not all preorders are preference orders, thinking of them as preference orders enables us to motivate the distinction between the pointwise product of $\mathcal{P} \times \mathcal{Q}$ of two preference orderings $\mathcal{P}$ and $\mathcal{Q}$ (which corresponds to impartiality between parameters $\mathcal{P}$ and $\mathcal{Q}$) and the lexicographic product (according to which any increase in $\mathcal{P}$ is more important than any increase in $\mathcal{Q}$). Naturally occurring preference orderings on products of posets tend to be complicated. Lexicographic products are extremely unlikely to represent your views on baskets of apples and oranges because even if you prefer apples to oranges, you would be unlikely to prefer any increase (however small) in the number of apples you are offered to any increase (however large) in the number of oranges, unless, that is, you have no use for oranges anyway. And in that case you would hardly prefer an apple and *two* oranges to an apple and *one* orange.

On the other hand, your preference ordering is likely nevertheless to be finer than the pointwise product ordering: according to the pointwise product ordering, you would be unable to decide between a single orange-with-a-pound-of-apples and two-oranges-with-one-apple. You would have to be very blasé not to prefer the first. After all, to a certain extent apples and oranges are interchangeable: realistic product (preference) orders refine the product order but are typically not as refined as a lexicographic order. We must not get too deeply into utility theory! Note merely that it is a sensible motivation for the study of orderings and products of orderings.

But before leaving preference orderings altogether the reader should notice at least that preference orders have a rather odd feature not shared by partial orders in general. $A \not\leq B \not\leq A$ and $B > C$ does not imply $A > C$, though one expects it to if the ordering is a preference ordering. This makes a nice exercise.

EXERCISE **3** *Are the two following conditions on partial orders equivalent?*

$(\forall xyz)(z < x \not\leq y \not\leq x \to z < y)$,
$(\forall xyz)(z > x \not\leq y \not\leq x \to z > y)$.

(This exercise uses three common conventions that it takes a logician to spell out. (i) When '$\leq$' and '$<$' appear in the same formula they denote a partial ordering and its strict part, respectively; (ii) that the relations $\leq$ and $\geq$ are converses of each other, and (iii) that '$x < y < z$' is short for '$(x < y) \wedge (y < z)$'.)

Given a subset $X \subseteq (P \times Q)$, the points in X that are maximal in the pointwise product $\mathcal{P} \times_{pw} \mathcal{Q}$ are called "**Pareto-efficient** points" by economists. They are sometimes called "Pareto-optimal" because if X is the set of points that are in some sense accessible, or possible, or something, then a Pareto-efficient point in X is one that, once one has reached it, one cannot find another point in X that makes one of the coordinates better without simultaneously making another one worse. Pareto was an Italian economist. Natural illustrations are defective in the way that we have seen that natural illustrations of lexicographic products are defective, but they might still help. The mathematician Green (after whom Green Street in Cambridge is named, and who invented Green functions) is the most famous most recent person of whom no picture survives. However, we will not develop these ideas here, as they find their most natural expression in connection with convex optimisation.

EXERCISE **4**

(i) *Show that a totally ordered poset is a lattice if and only if it has a top and bottom element. Show that such a poset is always distributive.*

(ii) *Which of the following are lattices?*
   (i) The set of subspaces of a vector space under inclusion.
   (ii) The set of positive integers under divisibility.
   (iii) The set of nonnegative integers under division.

(iv) *The set of square-free numbers under division.*
Where you find a lattice say whether or not it is distributive.

(iii) *Let X be an arbitrary infinite set. Discuss the following sets and explain whether or not they are lattices, complete lattices or chain-complete partial orders.*
(i) The set of all transitive relations on X, partially ordered by set inclusion.
(ii) The set of all total orderings of subsets of X, partially ordered by set inclusion.
(iii) The set of all antisymmetric relations on X, partially ordered by set inclusion.

(iv) *Show that distributivity and dual distributivity are the same (see definition page 16).*

(v) *Let $\mathcal{P} = \langle P, \leq_P \rangle$ and $\mathcal{Q} = \langle Q, \leq_Q \rangle$ be two posets. Are $\mathcal{P} \times_{lex} \mathcal{Q}$ and $\mathcal{Q} \times_{lex} \mathcal{P}$ isomorphic? Are the two (pointwise) products $\mathcal{P} \times \mathcal{Q}$ and $\mathcal{Q} \times \mathcal{P}$ isomorphic?*

(vi) *Give some examples to show that chain-complete posets are not always complete lattices.*

(vii) *Consider the set $I^2 = \{\langle x, y \rangle \in \Re^2 : 0 \leq x, y \leq 1\}$, the unit square in the first quadrant in the plane. Equip I^2 with the pointwise order. Identify the maximal elements (if any) and the sup of the following sets:*
(i) The points on the circle radius $1/2$ and centre $\langle 1/2, 1/2 \rangle$.
(ii) The points in the open disc radius $1/2$ and centre $\langle 1/2, 1/2 \rangle$.
(iii) The points with irrational coordinates in I^2.
Now do the same for the lexicographic order.

1.6 Logical connectives

We will use standard notation for the connectives of propositional logic: '$\vee$', '$\wedge$' for 'or' and 'and'. We will also write $\bigwedge_{i \in I} p_i$ and suchlike for indexed conjunctions (and disjunctions). We write '$p \to q$' for the connective that will be equivalent to '$\neg(p \wedge \neg q)$' or to '$\neg p \vee q$'. $\to$ is the **material conditional**. A conditional[1] is a binary connective that is an attempt to formalise a relation of implication. The two components glued together by the connective are the **antecedent** (from which one

---

[1] This word 'conditional' is overloaded as well. Often a formula whose principal ('top level') connective is a conditional will be said to be a conditional.

infers) and the **consequent** (which is what one infers). The material conditional is the simplest one: $p \to q$ evaluates to `true` unless p evaluates to `true` and q evaluates to `false`.

Lots of students dislike the material conditional as an account of implication. The usual cause of this unease is that in some cases a material conditional evaluates to `true` for what seem to them to be spurious and thoroughly unsatisfactory reasons: namely, that p is false or that q is true. How can q follow from p merely because q happens to be true? The meaning of p might have no bearing on q whatever! This unease shows that we think we are attempting to formalise a relation between *intensions* rather than a relation between *extensions*. $\land$ and $\lor$ are also relations between intensions but they also make sense applied to extensions. Now if p implies q, what does this tell us about what p and q evaluate to? Well, at the very least, it tells us that p cannot evaluate to `true` when q evaluates to `false`. This rule "from p and $p \to q$ infer q" is called **modus ponens**. q is the **conclusion**, p is the **minor premiss** and $p \to q$ is the **major premiss**. Thus we can expect the *extension* corresponding to a conditional to satisfy *modus ponens* at the very least.

How many extensions are there that satisfy *modus ponens*? It is easy to check (miniexercise) that the following list is exhaustive: $\lambda pq.q$, $\lambda pq.(p \longleftrightarrow q)$, $\lambda pq.\neg p$, $\lambda pq.(\neg p \lor q)$, $\lambda pq.$`false`. Evidently the material conditional is the *weakest* of these: the one that holds in the largest number of cases. To be precise: among those functions in $\{$`true`, `false`$\}^2 \to \{$`true`, `false`$\}$ that satisfy *modus ponens* it is the greatest in the sense of the ordering on maps from posets to posets that we defined on page 17: it is the conditional that evaluates to `true` whenever any conditional does.

We had better check that our policy of evaluating $p \to q$ to `true` unless there is a very good reason not to does not get us into trouble. Fortunately, in cases where the conditional is evaluated to `true` *merely* for spurious reasons, then no harm can be done by accepting that evaluation. For consider: if it is evaluated to `true` *merely* because p evaluates to `false`, then we are never going to be able to invoke it (as a major premiss at least), and if it is evaluated to `true` *merely* because q evaluates to `true`, then if we invoke it as a major premiss, the only thing we can conclude, namely q, is something we knew anyway.

This last paragraph is not intended to be a *justification* of our policy of using only the material conditional: it is merely intended to make it look less unnatural than it otherwise might. The astute reader who spotted that nothing was said there about conditionals as *minor* premisses should not complain. They may wish to ponder the reason for this omission.

Reasonable people might expect that what one has to do next is solve the problem of what the correct notion of conditional is for intensions. This is a very hard problem, since it involves thinking about the internal structure of intensions and nobody really has a clue about that. (This is connected to the fact that we do not really have robust criteria of identity for intensions, as mentioned earlier.) It has spawned a vast and inconclusive literature. Fortunately it turns out that we can duck it, and resolve just to use the material conditional all the time.

Before we leave conditionals altogether: the conditional $\neg B \to \neg A$ is the **contrapositive** of the conditional $A \to B$, and the **converse** is $B \to A$. (cf., converse of a relation). A formula like $A \longleftrightarrow B$ is a **biconditional**.

In *modus ponens* one *affirms* the antecedent and *infers* the consequent. *Modus tollens* is the rule:

$$\frac{A \to B \quad \neg B}{\neg A}.$$

Affirming the consequent and inferring the antecedent:

$$\frac{A \to B \quad B}{A}$$

is a **fallacy** (= defective inference). This is an important fallacy, for reasons that will emerge later. It is the **fallacy of affirming the consequent**.

2

Recursive datatypes

2.1 Recursive datatypes

2.1.1 Definition

'Recursive datatype' is the sexy, postmodern, techno-friendly way to talk about things that mathematicians used to call 'inductively defined sets'. I shall abbreviate these two words to the neologism 'rectype'.

The standard definition of the naturals is as the least set containing zero and closed under successor, or, using some notation we have just acquired:

$$\mathbb{N} = \bigcap \{Y : 0 \in Y \wedge S\text{``}Y \subseteq Y\}.$$

Of course $\mathbb{N}$ is merely the simplest example, but its definition exhibits the central features of a declaration of a rectype. In general, a rectype is a set defined as the smallest ($\subseteq$-least) set containing some **founders**[1] and closed under certain functions, commonly called **constructors**. (This *is* standard terminology.) $\mathbb{N}$ has only one founder, namely, 0, and only one constructor, namely, successor (often written 'S' or '`succ`': $S(x)$ is $x + 1$). For the record, a founder is of course a nullary (0-place) constructor.

2.1.2 Structural induction

This definition of $\mathbb{N}$ justifies induction over it. If $F(0)$ and $F(n) \rightarrow F(n+1)$, then $\{n : F(n)\}$ is one of these Y that contains 0 and is closed under S, and therefore it is a superset of $\mathbb{N}$, so every natural number is F. It is a bit like original sin: if F is a property that holds of 0,

---

[1] This is not standard terminology, but I like it and will use it.

and holds of $n + 1$ whenever it holds of n, then each natural number is innoculated with it as it is born. Hence induction.

It also justifies definition by recursion. You might like to try proving by mathematical induction that – for example – all functions satisfying the recursion

$$0! := 1; \ (n + 1)! := (n + 1) \cdot n!$$

agree on all arguments. That is to say we can use induction to prove the uniqueness of the function being defined.

2.1.3 Generalise from $\mathbb{N}$

$\mathbb{N}$ is of course the simplest example of a rectype: it has only one founder and only one constructor, and that constructor is unary.

My first encounter with rectypes was when I was exposed to compound past tenses in Latin, when I was about eight. I pointed out to my Latin teacher that the construction that gives rise to the pluperfect tense from the perfect (in "By the time I reached the station the train had left" the first verb is in the perfect and the second is in the pluperfect) could be applied again, and what was the resulting tense called, please? Maybe the reader has had similar experiences. In UK law, if it is a crime to do X, it is also a crime to attempt to do X or to conspire to do X. So presumably it is a crime to attempt to conspire to do X? Crimes and tenses form recursive datatypes.

The examples that will concern us here will be less bizarre. An X-**list** is either the empty object or the result of `consing` a member of X onto the front of an X-list. Thus a list can be thought of as a function from an initial segment of $\mathbb{N}$ to X. Thought of as a rectype, the family of X-lists has a founder (the empty list) and a single binary constructor: `cons`. Later in this book there will be illustrations using `ML` pseudocode, and in `ML` the notation '`h::t`' denotes the list obtained by `consing` the object `h` onto the front of the list `t`. `t` is the **tail** of `h::t`, and `h` is its **head**.

Rectypes are ubiquitous, and different tribes will find different examples obvious. Computer scientists might think of lists; mathematicians might think of the subgroup of a group generated by a set of elements of the group; a more advanced example is the family of Borel sets – one can prove things about all Borel sets by showing that every open set has a property F, the complement of a thing with F has F and the union of countably many things with F has F. Logicians will think of the rectype of formulæ or the rectype of primitive recursive functions that we will see later. Words in an algebra form a rectype. A bundle of important examples which we will discuss later features the transitive closure $^*R$ of a (binary) relation R, which is the intersection of all transitive supersets of R, and the symmetric closure of R (the intersection of all symmetric supersets of R) and the reflexive closure similarly.

The most appealing rectype of all is Conway Games (see Conway (2001)). Donald Knuth has popularised Conway's material in a book with the catchphrase "surreal numbers", but readers of this book should be equal to reading Conway's original.

We can develop analogues of mathematical induction for any recursive datatype, and I shall not spell out the details here, as we shall develop them in each case as we need them. This kind of induction over a rectype is nowadays called **structural induction**.[1]

2.1.4 Well-founded induction

2.1.4.1 Well-founded relations and induction

Suppose we have a carrier set with a binary relation R on it, and we want to be able to infer

$$\forall x \ \psi(x)$$

from

$$(\forall x)((\forall y)(R(y, x) \ \rightarrow \psi(y)) \rightarrow \psi(x)).$$

In words, we want to be able to infer that everything is ψ from the news that you are ψ as long as all your R-predecessors are ψ. **y is an R-predecessor of** x if $R(y, x)$. Notice that there is no "case $n = 0$" clause in this more general form of induction: the premiss we are going to use implies immediately that a thing with no R-predecessors must have ψ. The expression "$(\forall y)(R(y, x) \rightarrow \psi(y))$" is called the **induction hypothesis**. The first line says that if the induction hypothesis is

---

[1] Historical note: Russell and Whitehead called it **ancestral induction** because they called the transitive closure of a relation the **ancestral** of the relation. (This is because of the canonical example: the transitive closure of the parent-of relation is the ancestor-of relation.) I used their terminology for years – and I still think it is superior – but the battle for it has been lost; readers should not expect the word 'ancestral' to be widely understood any longer, though they may see it in the older literature.

However, in set theory 'transitive closure' is used to mean something different, and I shall continue to use 'ancestral' instead of 'transitive closure' where this is needed to preclude ambiguity.

satisfied, then x is ψ too. Finally, the inference we are trying to draw is this: **if** x has ψ whenever the induction hypothesis is satisfied, **then** everything has ψ. When can we do this? We must try to identify some condition on R that is equivalent to the assertion that this is a legitimate inference to draw in general (i.e., for any predicate ψ).

Why should anyone want to draw such an inference? The antecedent says "x is ψ as long as all the immediate R-predecessors of x are ψ", and there are plenty of situations where we wish to be able to argue in this way. Take $R(x, y)$ to be "x is a parent of y", and then the inference from "children of blue-eyed parents have blue eyes" to "everyone has blue eyes" is an instance of the rule schematised above. As it happens, this is a case where the relation R in question does *not* satisfy the necessary condition, for it is in fact the case that children of blue-eyed parents have blue eyes and yet not everyone is blue-eyed.

To find what the magic ingredient is, let us fix the relation R that we are interested in and suppose that the inference

$$\frac{(\forall y)(R(y, x) \to \psi(y)) \to \psi(x)}{(\forall x)(\psi(x))}$$

has failed for some choice ψ of predicate.[1] Then we will see what this tells us about R. To say that R is well-founded all we have to do is stipulate that this failure (whatever it is) cannot happen for any choice of ψ.

Let ψ be some predicate for which the inference fails. Consider the set of all things that are *not* ψ. Let x be something with no R-predecessors. Then all R-predecessors of x are ψ (vacuously!) and therefore x is ψ too. This tells us that if y is something that is not ψ, *then there must be some* y' *such that* $R(y', y)$ *and* y' *is not* ψ *either*. If there were not, y would be ψ. This tells us that the collection of things that are not ψ "has no R-least member" in the sense that everything in that collection has an R-predecessor in that collection.

Thus we can see that if induction fails over R, then there is a subset X of the carrier set (to wit, the extension of the predicate for which induction fails) such that every member of X has an R-predecessor in X.

One might have expected that for the inference to be good one would have had to impose conditions on both R and ψ. It is very striking that there should be a condition on R alone that is enough by itself for

---

[1] This is a common way of representing arguments in logic: premises above and conclusions below the line.

this inference to be good for *all* ψ. All we have to do is exclude the possibility of the domain of R having any such pathological subsets and we will have justified induction over R. Accordingly, we will attach great importance to the following condition on R:

DEFINITION 1 R *is* **well-founded** *iff every nonempty subset X of the domain of R has an element x such that all the R-predecessors of x lie outside X. (x is an "R-minimal" element of X.)*

This definition comes with a health warning: it is easy to misremember. The only reliable way to remember it correctly is to rerun in your mind the discussion we have gone through: well-foundedness is precisely what one needs a relation R to have if one is to be able to do induction over R. No more and no less. The definition is not memorable, but it is reconstructible.

A **well-ordering** is a well-founded strict total order. (No well-founded relation can be reflexive , so well-founded orders have to be of the strict flavour). Perhaps we should have some examples of well-orderings. Obviously any finite total order will be a well-order! What about infinite well-orderings? The only natural example of an infinite well-ordering is one we have already seen – $\langle \mathbb{N}, <_{\mathbb{N}} \rangle$. Notice that the real line $\langle \Re, <_{\Re} \rangle$ is not a well-ordering, for it is a simple matter to find sets of real numbers with no least element, for example, the set of all real numbers strictly greater than 0. This set has a lower bound all right, namely, 0 but this lower bound is not a member of the set and so cannot be the least member of it.[1]

Miniexercise: One can define well-orderings as relations that are trichotomous and well-founded.

We note here two facts that will come in useful later (see remark 39 and chapter 7):

EXERCISE 5 *A pointwise product of two well-founded (strict) partial orders is a well-founded (strict) partial order.*

A lexicographic product of two well-founded (strict) partial orders is a well-founded (strict) partial order.

It is not hard to see that for a finite binary structure to be well-

---

[1] It is important not to get confused (as many people do) by the fact that every set of reals has a *greatest lower bound*. For example, $\{x \in \Re : x > 0\}$ has no least member, but it does have a greatest lower bound, which is of course 0. Notice that $0 \notin \{x \in \Re : x > 0\}$!!

founded it is neccessary and sufficient for it to have no loops: a loop is manifestly a subset with no least element! This necessity remains with infinite structures, but it is no longer sufficient: the negative integers with the relation $\{\langle n, n-1 \rangle : n \in Z^- \}$ has no loops, but it is still not well-founded. With the help of an apparently minor assumption we can show that this is the only badness that can happen in infinite ill-founded structures. The **Axiom of Dependent Choices**, usually known as DC, says that if R is a relation such that $(\forall x \in Dom(R))(\exists y)(R(x,y))$, then there is an infinite R-chain.

For the moment a (well-founded) **tree** is a poset with a bottom element where for every element x the set $\{y : y < x\}$ is a well-ordering. It is **finite branching** iff each point has at most finitely many immediate successors. A weak version of DC that has many uses is **König's Infinity Lemma**:

KL: Every finite branching tree with infinitely many nodes has an infinite path.

EXERCISE 6

(i) *Use DC to prove that for every bounded set $X \subseteq \Re$ there is an increasing sequence $\langle x_i : i \in \mathbb{N} \rangle$ of elements of X with the same sup as X.*

(ii) *Let us say an (n, d) tree is one where every node has at most d children, and no path is of length $> n$. Show that such a tree has only finitely many nodes. Why does this not imply König's Infinity Lemma?*

The full **Axiom of Choice**, which we will need later, is

If X is a set of nonempty sets, there is a function $f : X \to \bigcup X$ s.t. $(\forall x \in X)(f(x) \in x)$.

Such a function is a **selection function**.

Now by use of DC we can show that any ill-founded structure admits an "infinite descending sequence". Loops are simply infinite descending sequences that happen to be periodic. So, assuming DC we could define: R is well-founded if there is no $f : \mathbb{N} \to dom(R)$ such that $(\forall n)(R(f(n+1), f(n))$.

Beware! One might think that the easiest and most natural definition of well-foundedness is this last one in terms of descending chains. (It is certainly a lot easier to understand!) However, defining well-foundedness in terms of descending chains means that one needs DC to deduce induction. Although DC might look plausible (and hardly anybody engaged in axiomatic mathematics shrinks from assuming it) it is a nontrivial assumption and it is good practice to prefer proofs that avoid it when such proofs are available.

The official definition of well-foundedness is a lot more unwieldy than the definition in terms of descending sequences. In consequence, it is very easy to misremember it. A common mistake is to think that a relation is well-founded as long as its domain has a minimal element, and to forget that *every* nonempty subset must have a minimal element. The only context in which this definition makes any sense at all is induction, and the only way to understand the definition or to reconstruct it is to remember that it is cooked up precisely to justify induction. This last fact is the content of the next theorem.

THEOREM 2 *R is a well-founded relation iff we can do well-founded induction over the domain of R.*

Proof: The left-to-right inference is immediate: the right-to-left inference is rather more interesting.

What we have to do is use R-induction to prove that every subset of the domain of R has an R-minimal element. But how can we do this by R-induction? The trick is to prove by R-induction ("on x") that every subset of the domain of R to which x belongs contains an R-minimal element. Let us abbreviate this to "x is R-**regular**".

Now let x_0 be such that every R-predecessor of it is R-regular, but such that it itself is not R-regular. We will derive a contradiction. Then there is some $X \subseteq dom(R)$ such that $x_0 \in X$ and X has no R-minimal element. In particular, x_0 is not an R-minimal element of X. So there must be x_1 s.t. $R(x_1, x_0)$ and $x_1 \in X$. But then x_1 is likewise not R-regular. But by hypothesis everything R-related to x_0 was R-regular, which is a contradiction.

Therefore everything in $dom(R)$ is R-regular. Now to show that any subset X of $dom(R)$ is either empty or has an R-minimal element. If X is empty, we are all right. If it is not, it has a member x. Now we have just shown by R-induction that x is R-regular, so X has an R-minimal element as desired. ∎

Well-foundedness is a very important concept throughout mathematics, but it is usually spelled out only by logicians. (That is why you read it here first.) Although the rhetoric of mathematics usually presents mathematics as a static edifice, mathematicians do in fact think dynamically, and this becomes apparent in mathematical slang. Mathematicians often speak of *constructions* underlying proofs, and typically for a proof to succeed it is necessary for the construction in question to terminate. This need is most obvious in computer science, where one routinely has the task of showing that a program is well-behaved in the sense that every run of it halts. Typically a program has a main loop that it goes through a number (which one hopes will be finite!) of times. The way to prove that it eventually halts is to find a parameter changed by passage through the loop. A common and trivial example is the `count` variable to be found in many programs that is not affected by the passage through the loop but only by the decrement command at the start of each pass. Sometimes the rôle is played by a program variable that is decremented at each pass – not explicitly decremented at the start of each pass like a `count` variable, but as a side-effect of what happens on each pass. In general we look for a parameter that need not be a program variable at all, but merely some construct put together from program variables that takes values in a set X with a binary relation R on it such that (i) at each pass through the loop the value of the parameter changes from its old value v to a new value v' such that $\langle v, v' \rangle \in R$ and (ii) any sequence v_0, v_1 ... where, for all n, $\langle v_n, v_{n+1} \rangle$ *in R is finite.*[1]

If we can do this, then we know that we can only make finitely many passes through the loop, so the program will halt. Condition (ii) is the descending-sequence version of well-foundedness.

EXERCISE 7 *The game of Sylver Coinage was invented by Conway, Berlekamp and Guy (1982). It is played by two players, I and II, who move alternately, with I starting. They choose natural numbers greater than 1 and at each stage the player whose turn it is to play must play a number that is not a sum of multiples of any of the numbers chosen so far. The last player loses.*

Notice that by 'sum of multiples' we mean 'sum of positive multiples'. The give-away is in the name: 'Sylver Coinage'. What the players are doing is trying at each stage to invent a new denomination of coin, one that is of a value that cannot be represented by assembling coins of the

---

[1] If you were expecting this sentence to end "is eventually constant", look ahead to section 2.1.4.3, p. 36.

denominations invented so far. (There is a significance to the spelling of 'silver', but I do not think we need to concern ourselves with that.)

Prove that no play of this game can go on forever.

The way to do this is to identify a parameter which is altered somehow by each move. The set of values that this parameter can take is to have a well-founded relation defined on it, and each move changes the value of the parameter to a new value related to the old by the well-founded relation. The question for you is, what is this parameter? and what is the well-founded relation?

(You should give a much more rigorous proof of this than of your answer to exercise 10 below: it is quite easy to persuade oneself that all plays are indeed finite as claimed, but rather harder to present this intuition as reasoning about a well-founded relation.)

As we noted earlier, we can think of binary relations as matrices, but we can also think of them as digraphs, where there is a vertex for each element of the domain and an edge from a to b if a is related to b. This is a very natural thing to do in the present context, since we can also think of the arrows as representing a possible step taken by the program in question. It also gives us a convenient way of thinking about composition and transitive closures. a is related to b by R^n if there is a path of length n from a to b in the digraph picture of R, and a is related to b by the transitive closure of R if there is a path from a to b at all. It also makes it very easy to see that the transitive closure of a symmetric relation is symmetric, and makes it obvious that every subset of a well-founded relation is well-founded. This makes it easy to explain why pointwise products of well-founded relations are well-founded.

The digraph picture gives rise to **Hasse diagrams**. When drawing a digraph of a transitive relation R one can safely leave out a lot of arrows and still display the same information: all one has to draw is the arrows for a relation whose transitive closure is R. Thus the relation represented by a lot of dots joined by arrows is the relation "I can get from x to y by following arrows".

EXERCISE 8 *Find an example to illustrate the fact that, for an arbitrary transitive relation, there is no minimal relation of which it is the transitive closure.*

In fact we can leave out the heads on the arrows (so we draw in edges rather than arrows) by adopting the convention that the end of the edge on which the arrowhead belongs is the end that is further up the page. (Of course this only works if the relation is transitive!) The result of doing this is the Hasse diagram of that transitive relation. The appeal of Hasse diagrams relies on and to some extent reinforces an unspoken (and false!) assumption that every partial order can be embedded somehow in the plane: every ascending chain is a countable linear order (in which the rationals cannot be embedded), and every antichain is isomorphic to a nowhere dense subset of $\Re$. Related to this is the weaker (but nevertheless still nontrivial) assumption that all total orders can be embedded in the real line, as instance, the image of Justice, blindfolded with a pair of weighing scales. Although this is clearly a false assumption that might perhaps push our intuitions in wrong directions – we in fact need a weak version of the axiom of choice (see exercise 29) to show that every partial order has a superset that is a total order – it is not such a crazy idea in computer science, where linearity of time and of machine addresses compel us to think about extensions of partial orders of precisely this kind.

2.1.4.2 Recursion on a well-founded relation

THEOREM 3 *Let $\langle X, R \rangle$ be a well-founded structure and $g : X \times V \to V$ be an arbitrary (total) function. Then there is a unique total function $f : X \to V$ satisfying $(\forall x \in X)(f(x) = g(x, f``\{y : R(y, x)\}))$*

Here V is the universe, so that when we say "$g : X \times V \to V$" we mean only that we are not putting any constraints on what the values of g (or its second inputs) are to be.

Proof: The idea is very simple. We prove by R-induction that for every $x \in X$ there is a unique function f_x satisfying $(\forall y)(^*R(y, x) \to f_x(y) = g(y, f_x``\{z : R(z, y)\}))$. We then argue that, if we take the union of the f_x, the result will be a function, and this function is the function we want. ∎

The following commutative diagram might help.

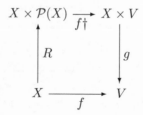

$f\dagger$ is $\lambda a.\langle \text{fst } a, f\text{“}(\text{snd } a)\rangle$ ("leave the first component alone and translate the second under f"). The map R is not the map from X into $\mathcal{P}(X)$ corresponding to R (remember that every subset of $X \times X$ corresponds to a map $X \to \mathcal{P}(X)$) but instead the map that sends a pair $\langle x, y \rangle$ to $\langle x, \{z : R(z, y)\}\rangle$. ($V$ contains everything: not just junk but sets of junk as well, so you don't have to worry about whether values of g are sets or junk.)

The reason this crops up here is that all rectypes – since they are generated by functions – will have a sort of **engendering relation**[1] that is related to the functions that generate the recursive datatype rather in the way that $<_{\mathbb{N}}$ is related to the successor function. The engendering relation is that binary relation that holds between an object x in the rectype and those objects "earlier" in the rectype out of which x was built. Thus it holds between a formula and its subformulæ, between a natural number and its predecessors and so on. Put formally, the (graph of the) engendering relation is the transitive closure of the union of the (graphs of the) constructors.

The (graph of, extension of) the engendering relation is itself a rectype. For example, $<_{\mathbb{N}}$ is the smallest set of ordered pairs containing all pairs $\langle 0, n \rangle$ with $n > 0$ and closed under the function that applies S to both elements of a pair (i.e., $\lambda p.\langle S(\text{fst } p), S(\text{snd } p)\rangle$).

The following triviality is important.

THEOREM 4 *The engendering relation of a rectype is well-founded.*

Proof: Let X be a subset of the rectype that has no minimal element in the sense of $<$, the engendering relation. We then prove by structural induction ("on x") that $(\forall y)(y < x \to y \notin X)$. ∎

We have not made the assumption that the constructors have finite arity, as we can prove that the engendering relation is well-founded even without it. There are several important examples of rectypes whose constructors do not have finite arity: one important example that we shall see later is the rectype of well-founded sets in set theory; a standard example from analysis is the rectype of Borel sets of reals, but by far the most attractive is the rectype of Conway games in (2001).

For the moment we will be concerned mainly with rectypes whose constructors are all of finite arity. Such a rectype will be said to have **finite character**. If in addition it has only finitely many of them, and

---

[1] This is not standard terminology.

only finitely many founders, it will be said to be **finitely presented**; if it has finitely many constructors of finite arity, and only countably many founders, it will be said to be **countably presented**.

Engendering relations on rectypes give us nice concepts of **bounded quantifiers**. The commonest, most natural and most important example is of bounded quantifiers is to be found in arithmetic: $(\forall n < m)(\ldots)$ and $(\exists n < m)(\ldots)$, which we will see a lot more of in chapter 6. The idea is that expressions with bounded quantifiers only – and no unbounded quantifiers – should be thought of as being quantifier-free. This is because an injunction to search for something $< x$ is only an injunction to search among things that you are given if you are given x, not to scour the entire universe.

(If this is puzzling to you because you do not feel comfortable with predicate languages, fear not. Put it on the back burner and return to it later, in chapter 5.)

Theorem 4 means that we can always do well-founded induction over the engendering relation. In this simplest case, $\mathbb{N}$, this well-founded induction is often called *strong* induction or sometimes *course of values* induction. Quite often arguments by well-founded induction are presented in contrapositive form (see page 22). We first establish that, if there is a counterexample to what we are trying to prove, then there is an earlier counterexample. So the set of counterexamples has no least element and so by well-foundedness must be empty. The standard example of this style of proof is due to Fermat, who proved that $x^4 + y^4 = z^2$ has no nontrivial solutions in $\mathbb{N}$. It uses the fact that all pythagorean triples are of the form $a^2 - b^2$, $2ab$, $a^2 + b^2$ to show that for any solution to $x^4 + y^4 = z^2$ there is one with smaller z. This gives us a proof by well-founded induction on $<_{\mathbb{N}}$ that there are no solutions at all. The details are fiddly, which is why it is not an exercise. The following are more straightforward.

EXERCISE 9 *Dress up the traditional proof that $\sqrt{2}$ is irrational into a proof by well-founded induction on $\mathbb{N} \times \mathbb{N}$.*

The following example is the most natural use of this technique known to me.

EXERCISE 10 *A square can be dissected into finitely many squares all of different sizes (see Gardner, 1961 chapter 17). Prove that a cube cannot be dissected into finitely many cubes all of different sizes.*

(Do not attempt to give too rigorous a proof.)

EXERCISE 11 *Show that the relation $\trianglelefteq$ on $\mathbb{N}$ defined by $n \trianglelefteq m$ iff $n < m \leq 100$ is well-founded.*

Consider the two functions f and g defined thus:

$f(x) = $ if $x > 100$ then $(x - 10)$ else $f(f(x + 11))$;

$g(x) = $ if $x > 100$ then $(x - 10)$ else 91.

Prove that f and g are the same functions-in-extension.

That was question 1993:9:10 from the Computer Science tripos at Cambridge, available at:

http://www.cl.cam.ac.uk/tripos/y1993.html

EXERCISE 12 *You presumably know the proof that the arithmetic mean of two reals is at least as big as the geometric mean. In fact this works for the arithmetic and geometric mean of any finite number of reals. The standard proof proceeds by showing that it works for two reals; and that if it works for n reals. it works for 2n reals; and that if it works for n reals, it works for $n - 1$ reals (see Aigner and Ziegler 2001, pp 99-100). This is a well-founded induction over a kinky relation on $\mathbb{N}$. What is this relation, precisely?*

Before we leave recursion on a well-founded relation it might be helpful to have an illustration. Discrete games in which all plays are finite and no draws are allowed always have a winning strategy for one player or another. Let us prove this.

Let X be an arbitrary set. (X is intended to be the set of *moves*.) $[X]^{<\omega}$ is the set of finite sequences of members of X. Let G be a subset of $[X]^{<\omega}$ **closed under shortening** (i.e., initial segments of sequences in G are also in G). There is a map v defined on the **endpoints** of G (sequences in G with no proper end-extensions in G) taking values in the set $\{\text{I}, \text{II}\}$.

Players I and II play a game by picking elements of X alternately, with I playing first, with their choices constrained so that at each finite stage they have built a finite sequence in G.

If they reach an endpoint of G, the game is over, and v tells them who has won. For the purposes of illustration we will assume that all plays in G are finite. (We shall see later how this assumption can be weakened: see exercise 19). The connection with well-foundedness is that this condition is captured by saying that the relation "$s \in G \wedge t \in G$ and s is an end-extension of t" is well-founded. (If you cannot work out which way round to read this, just note: one way round it is *obviously* well-founded: what we mean is that it is well-founded the other way round too.)

Next we need the notion of **even** and **odd** positions. A sequence from X of even length is a position when it is I's turn to move; a sequence from X of odd length is a position when it is II's turn to move. Clearly, if s is an even position and even one of its children (positions to which I can move at his next move) is labelled 'I', then we can label s 'I' too, since I can win from there. Similarly, if s is an *odd* position and *all* its children (positions to which II can move at his next move) are labelled 'I', then s can be labelled 'I' too. This ratcheting up the upside-down tree of finite sequences that comprise G is a recursive definition of a labelling extending v that – because of well-foundedness – is defined on the whole of G. Thus the empty sequence ends up being labelled, and the lucky owner of the label has a winning strategy.

It is very important that no assumption has been made that X is finite, nor that there is a finite bound on the length of sequences in G.

2.1.4.3 Other definitions of well-foundedness

It is clearly an immediate consequence of our definition of well-foundedness that any well-founded relation must be irreflexive. Nevertheless, one could define a relation $R \subseteq X \times X$ to be well-founded if $(\forall X' \subseteq X)(\exists x \in X')(\forall x' \in X')(R(x', x) \rightarrow x = x')$. This definition of well-foundedness has a "descending chain" version too: "every R-chain is eventually constant". This definition is more appealing to some tastes. It has the added advantage over the other definition that it distinguishes between a well-ordering of the empty set (which will be the empty relation) and a well-ordering of the singleton $\{x\}$, which will be the relation $\{\langle x, x \rangle\}$. In contrast, according to the other definition, the empty relation is not only a well-ordering of the empty set, but it is also a well-ordering of the singleton $\{x\}$!

It is a miniexercise to verify that each concept of well-foundedness is definable in terms of the other. The situation is rather like that with regard to strict and nonstrict partial orders.

2.1.4.4 Structural induction again

We know that structural induction holds for rectypes, but we could deduce it from the well-foundedness of the engendering relation if we wished. Take the example of $\mathbb{N}$. Suppose we know that 0 has property F, and that whenever n has property F so does $S(n)$. Then the set of integers that are *not* F (if there are any) will have no least member and therefore, by well-foundedness of $<_{\mathbb{N}}$, will be empty.

This holds in general: we can deduce structural induction from the well-foundedness of the engendering relation. For example, if we can prove $(\forall n)(\Phi(n))$ by a well-founded induction over $<_{\mathbb{N}}$, then we can prove $(\forall n)(\forall m <_{\mathbb{N}} n)(\Phi(m))$ by structural induction.

Recursive datatypes ⟶ Structural induction

Well-foundedness of engendering relation → well-founded induction

2.1.4.5 Other uses of well-foundedness

Intuitions of well-foundedness and failure of well-foundedness are deeply rooted in common understandings of impossibilities. For example: it is probably not unduly fanciful to claim that the song "There's a hole in my bucket, dear Liza" captures the important triviality that a process that eventually calls itself with its original parameters will never terminate. The attraction of tricks like the ship-in-a-bottle seems to depend on the illusion that two processes, each of which (apparently) cannot run until it has successfully called the other, have nevertheless been successfully run. A similar intuition is at work in the argument sometimes used by radical feminists to argue that they can have no (nonsexist) surnames, because if they try to take their *mother's* surname instead of their fathers, then they are merely taking their *grandfather's* surname, and so on. Similarly one hears it argued that, since one cannot blame the person from whom one catches a cold for being the agent of infection (for if one could, they in turn would be able to pass the blame on to whoever infected them, and the process would be ill-founded[1]), so one cannot blame anyone at all. This argument is used by staff in STD clinics to help their patients overcome guilt feelings about their afflictions.

The reader is invited to consider and discuss the following examples from the philosophical literature.

(i) "...most of those who believe in probability logic uphold the view that the appraisal is arrived at by means of a principle of induction[2] which ascribes probabilities to the induced hypothesis. But

---

[1] Unless one can blame Eve!

[2] This is of course philosophical not mathematical induction!

if they ascribe a probability to this 'principle of induction' in turn, the infinite regress continues".

Popper (1968) p 264.

(ii) "In every judgement, which we can form concerning probability, as well as concerning knowledge, we ought always to correct the first judgement, deriv'd from the nature of the object, by another judgement, deriv'd from the nature of the understanding. 'Tis certain a man of solid sense and long experience ought to have, and usually has, a greater assurance in his opinions, than one who is foolish and ignorant, and that our sentiments have different degrees of authority, even with ourselves, in proportion to the degrees of our reason and experience. In the man of the best sense and longest experience, this authority is never entire; since even such-a-one must be conscious of many errors in the past, and must still dread the like for the future. Here then arises a new species of probability to correct and regulate the first, and fix its just standard and proportion. As demonstration is subject to the control of probability, so is probability liable to a new correction by a reflex act of the mind, wherein the nature of our understanding, and our reasoning from the first probability become our subjects.

"Having thus found in every probability, beside the original uncertainty inherent in the subject, a new uncertainty deriv'd from the weakness of that faculty, which judges, and having adjusted these two together, we are oblig'd by our reason to add a new doubt deriv'd from the possibility of error in the estimation we make of the truth and fidelity of our faculties. This is a doubt, which immediately occurs to us, and of which, if we wou'd closely pursue our reason, we cannot avoid giving a decision. But this decision, though it shou'd be favourable to our preceding judgement, being founded only on probability, must weaken still further our first evidence, and must itself be weaken'd by a fourth doubt of the same kind and so *ad infinitum*; till at last there remain nothing of the original probability, however great we may suppose it to have been, and however small the diminution by every new uncertainty. No finite object can subsist under a decrease repeated *in infinitum*; and even the vastest quantity, which can enter into human imagination, must in this manner be reduc'd to nothing."

Hume (1739) book I part IV, sec 1, 5-6.

(Miniexercise: Prove Hume wrong by exhibiting an infinite family of numbers strictly between 0 and 1 whose product is nonzero.)

(iii) "Volitions we postulated to be that which makes actions voluntary, resolute [etc.]. But ... a thinker may ratiocinate resolutely, or imagine wickedly .... Some mental processes then can, according to the theory, issue from volitions. So what of the volitions themselves? Are they voluntary or involuntary acts of mind? Clearly either answer leads to absurdities. If I cannot help willing to pull the trigger, it would be absurd to describe my pulling it as voluntary. But if my volition to pull the trigger is voluntary, in the sense assumed by the theory, then it must issue from a prior volition and from that another ad infinitum."

Ryle (1983) pp. 65-6.

2.1.5 Sensitivity to set existence

We now return to structural induction and consider how the set formulæ for which one can perform structural induction over a rectype depends on what other assumptions one makes. We deduce an instance of structural induction over a rectype by appealing to the fact that the rectype (of widgets, as it might be) is the intersection of all things containing the founders and closed under the constructors. So if the class of things that are F contains the founders and is closed under the constructors, then all widgets are F. For this to work we need to know that the extension (page 7) of F really exists. In this way we see that the extent of what we can prove by induction is determined at least in part by our set existence axioms. This will matter later on when we start doing set theory.

2.1.6 Countably presented rectypes are countable

Mathematicians should be warned that logicians often use the word 'countable' to mean 'countably infinite'. The symbol used for the cardinal number of countably infinite sets is '$\aleph_0$'.

2.1.6.1 The prime powers trick

"Countably presented" is slang, but in this case we mean that the rectype has countably many founders and countably many functions all of finite arity.

THEOREM 5 *Countably presented rectypes are countable.*

Sketch of proof: The key observation is that the set of finite sets of naturals and the set of finite sequences of naturals are both countable. The function $\lambda x.\Sigma_{n \in x} 2^n$ maps finite sets of natural numbers 1-1 to natural numbers. (Make a note of this for later use in finding models of ZF without the axiom of infinity.) We map finite sequences of naturals to naturals by sending – for example – the tuple $\langle 1, 8, 7, 3 \rangle$ to $2^{1+1} \cdot 3^{8+1} \cdot 5^{7+1} \cdot 7^{3+1}$). This is the **prime powers trick**.

The elements of a finitely presented (indeed countably presented) rectype can obviously be represented by finite sequences of symbols, and so the prime powers trick is enough to show that every finitely presented rectype is countable.   ∎

(The reason that this is only a sketch is that we are not considering at this stage how to establish that a declaration of a rectype succeeds in creating a set at all. We will take this up in chapter 8.) I shall say nothing at all at this stage about how big a rectype can be if it is not countably presented.

A meal is often made of the fact that (the syntax of) every human language is a rectype – unlike the syntax of any animal language – and that therefore the repertory of possible expressions is infinite in a way that the repertory of meaningful calls available to animals of other species is not. Quite how useful this recursive structure is to those who wish to drive a wedge between human language and animal language is not entirely clear, but it is immensely useful when dealing with artificial languages, since it enables us to exploit structural induction in proving facts about them.

We can enumerate the wffs[1] of a language and then *sequences* of wffs of a language (which is to say, Gödel-proofs, which we will meet on page 76). This enables us to arithmetise proof theory and eventually to prove the incompleteness theorem. Since this was first done by Gödel with precisely this end in view, any enumeration of formulæ (or register machines or Turing machines, as in chapter 6, or anything else for that matter) tends to be called **Gödel numbering** or **gnumbering** for short: the 'g' is silent.)

The way to crystallise the information contained in theorem 5 is to develop a nose for the difference between what one might call *finite precision objects* and *infinite precision objects*. Members of finitely presented

---

[1] Logicians' slang, which I shall frequently lapse into. It is an acronym: Well-Formed Formula.

rectypes are finite precision objects: one can specify them uniformly with only finitely many symbols. In contrast, the reals, for example, are infinite precision objects: there is no way of uniformly notating reals using only finitely many symbols for each real. There is a sort of converse to theorem 5: if a set is countable, then there will be a way of thinking of it (or at least there will be a notation for its members) as a finitely presented rectype. This gives us a rule of thumb: if X is a set that admits a uniform notation for its members, where each member has a finite label, then X is countable, and conversely. This is not the *definition* of countable, but it is the most useful way to tell countable sets from uncountable sets.

EXERCISE 13 *Which of the following sets are countable and which are uncountable, using the above test? The set of*
(i) permutations of IN *that move only finitely many things;*
(ii) permutations of IN *of finite order;*
(iii) algebraic numbers;
(iv) partitions of IN *into finitely many pieces;*
(v) partitions of IN *all of whose pieces are finite;*
(vi) partitions of IN *containing a cofinite piece;*
(vii) increasing functions IN → IN*;*
(viii) nonincreasing functions IN → IN $(f(n+1) \leq f(n))$.

It is often quite hard to provide explicit bijections between two things that are the same size: a good example is the naturals and the rationals. However, it commonly happens that we find sets X and Y where – although there is no obvious bijection between them – there are nevertheless obvious injections $X \hookrightarrow Y$ and $Y \hookrightarrow X$. For this reason we often have recourse to the Schröder-Bernstein theorem (theorem 3.1.1), which tells us that in those cicumstances there is actually a bijection between X and Y after all, and it can even be given explicitly if we are patient.

However, proving that the uncountable sets mentioned in this exercise are all the same size is quite hard even with the aid of theorem 3.1.1!

EXERCISE 14 *Exhibit injections from the rationals into the naturals, and vice versa.*

2.1.6.2 Cantor's theorem

Although we will not need this until later, we may as well note at this stage that there are precisely as many countable sequences of reals as there are reals. (To show there are as many countable *sets* of reals as reals one needs countable choice (page 61).)

If one rolls a die and tosses a coin, there are $6 \times 2 = 12$ possible outcomes: "multiply independent possibilities". This principle works in general, and not just when the the number of factors or the number of outcomes at each trial is finite. So how many real number are there? Every real has a binary representation, which is $\aleph_0$ independent choices from $\{0,1\}$, so there must be $2^{\aleph_0}$ reals. Well, almost! There is an unwelcome complication in that rationals whose denominators are powers of 2 have two representations in this form, and we will see later (lemma 91) how to iron out this wrinkle. Slightly less elegant, but more immediately successful, is the idea of using continued fraction representations, and this of course will tell us by the same token that there are precisely $\aleph_0^{\aleph_0}$ reals. In fact, $\aleph_0^{\aleph_0} = 2^{\aleph_0}$, but it takes a certain amount of tinkering to prove it (it is a miniexercise and uses theorem 3.1.1) and sadly there is no very easy proof that $|\Re| = 2^{\aleph_0}$: all proofs involve either lemma 91 or the fact that $\aleph_0^{\aleph_0} = 2^{\aleph_0}$.

EXERCISE 15 *Find a bijection between* $\mathbb{N} \times \mathbb{N}$ *and* $\mathbb{N}$. *Use it and the result of the miniexercise to show that there are precisely as many* ω *sequences of reals (sequences indexed by* $\mathbb{N}$*) as there are reals.*

The finite-description test shows that the set of finite subsets of $\mathbb{N}$ is countable. In contrast, the set of *all* subsets of $\mathbb{N}$ is not. This is a consequence of the following theorem, which is easy and of central importance.

THEOREM 6 *(Cantor's theorem) There is no surjection from any set onto its power set.*

Proof: Let f be a map from X to $\mathcal{P}(X)$. We shall show that f is not onto. Let $C = \{x \in X : x \notin f(x)\}$. If f were onto, we would have $C = f(a)$ for some $a \in X$. But then we can reason as follows. $a \in f(a)$ iff $a \in C$ (since $C = f(a)$) iff $a \notin f(a)$ (by membership condition on C), whence $a \in C \longleftrightarrow a \notin C$.   ∎

"Multiply independent probabilities" tells us that $|\mathcal{P}(X)| = 2^{|X|}$. This means that we have proved that $x < 2^x$ for all cardinal numbers x. And it really is all cardinal numbers, even the infinite ones, because we have nowhere assumed that X was a finite set. It is also worth noting that the proof is constructive: not only does it show that no $f : X \to \mathcal{P}(X)$

can be onto, it also embodies an algorithm that for each $f : X \to \mathcal{P}(X)$ exhibits a subset of X not in the range of f.

2.1.7 Proofs

One last general point about rectypes that we should note is the idea of a *proof* (that something is in a rectype). For the moment let us restrict our attention to rectypes of finite character.

The idea is that if something turns out to be in a particular rectype, then there is a good finite reason for it to be, such as a construction of the object by means of the functions the rectype is built with. Thus $4 \in \mathbb{N}$ because of $\{0, 1, 2, 3, 4\}$. More generally, if R is the engendering relation of the rectype, then $R``\{x\}$ is a proof that $x$ is in the rectype: $R``\{x\}$ contains everything that needs to be checked to confirm x's membership.

Perhaps a better way of putting this would be to say that $\{0, 1, 2, 3, 4\}$ is a *manifestation* of the natural-numberhood of 4, but these constructions are increasingly coming to be called **proofs**. (Some communities use the word *'certificate'* in contexts like this: a pair of factors is a *certificate* for the compositeness of a number.) This might look rather like a loose usage of an old word, but the circle will close when we show that formal concepts of proofs (at least since Gödel, as we shall see on page 76) have in fact been constructions of this kind.

2.2 Languages

Important examples of rectypes for us in logic are *languages*, and it is to these that we now turn. An **alphabet** is a (usually but not invariably finite) set of atomic symbols, like the alphabet a, b, c ..., typically written in the style: {a,b,c}. A **string** or a **word** (or **formula** in most of the languages of interest to us) is a (for these purposes) finite list (or sequence) of letters from an alphabet. If Σ is an alphabet, this set of all finite sequences from Σ is often called 'Σ^*'. A **language** is a set of words (or strings or formulæ). Notice that this is an entirely syntactic definition. The semantics will come later.

There is one family of languages that we will not be concerned greatly with but which makes a useful way in to languages we do need. These are the so-called *regular languages*.

Consider the machine in figure 2.1. If this machine is started in the designated start state, and moved by inputs from one state to another according to the labels on the arrows, we can see that it will be in a

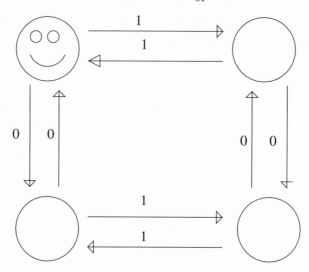

Fig. 2.1. A finite-state machine

state labelled by a smiley iff it has received an even number of 0's and an even number of 1's. We say that this machine **accepts** strings having an even number of 0's and an even number of 1's and that it **recognises** the set of strings having an even number of 0s and and even number of 1's. This is a common cause of confusion: the machine *accepts* strings, but it *recognises* sets of strings.

Machines that can be drawn in this style are **finite-state** machines. They are not constrained to have only one smiley (though they can have only one start state!) and the start state need not have a smiley. Such a machine knows only which state it is in. It does not know which state it was in last, and it does not know how often it has been in any given state. We say of a set of strings that is recognised by one of these machines that it is a **regular language**.

At this stage the only useful example of a regular language is the positional notation for natural numbers to base n for fixed n. In figure 2.2 is a finite-state machine that accepts strings of 0's and 1's that start with a 1 and thereby recognises the set of binary representations of members of $\mathbb{N}^+$.

This machine accepts any string of 0's and 1's beginning with a 1. This set of strings is the set of base-2 notations for natural numbers and is thus a set of **numeral**s. (Remember to distinguish between a natural number and a notation for it.)

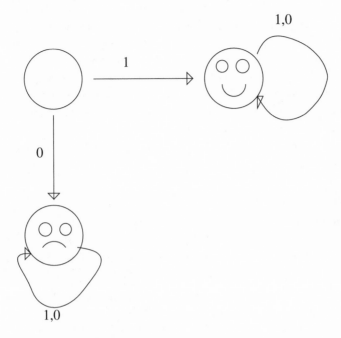

Fig. 2.2. A finite-state machine recognising numerals to base 2

The scowly is not standard but I use it. The smiley is not standard either: the official symbol is a pair of concentric circles.

It has to be admitted that the use of the word 'language' in this connection is a bit question-begging: not all regular languages have any semantics. The set of strings of 0's and 1's starting with a 1 is an example of a regular language with a natural semantics, but there are not many. The older terminology, now obsolescent, speaks of regular *events* and not of regular *languages*.

It is now possible to see why regular languages are not going to be of much use to us, for consider the set of strings of left-and-right brackets where every left bracket is closed by a right bracket and there are no extra right brackets (the "matching brackets" language). Any machine that accepts strings of matching brackets and accepts no other strings must be able to keep track of how many left brackets have been opened, and this can get arbitrarily large, and therefore larger than the number of states of any given machine. In a way this is unfortunate, since any language that is going to admit nontrivial semantics is certainly going to have at least the complexity of the matching-bracket language. Those that might appear not to, like Polish notation, have the same complexity in a less obvious manner.

There are natural examples of regular languages outside mathematics: it seems that for every natural language the sound strings that form permissible words of that language constitute a regular language.

Regular languages will have little further application in this text, but here they have served to introduce us to machines, and ideas of *accepting* and *recognising*, which we will need in connection with Turing machines in chapter 6.

2.2.0.1 Ambiguous parsing

Not surprisingly, we are going to be interested only in languages that can be used to say things. In practice this means languages that are rectypes – and not even all of them. Unless the recursive datatype that is the language is in some sense free (so that each object in it can be generated in only one way – like ℕ) some strings will turn up in more than one way. This means that one regards the formula not just as a string but as a string with extra structure that tells us where the string comes from, such as a *proof* or *certificate* in the sense of section 2.1.7.

To the recursions that generate the strings of which the language is composed will correspond rules telling us how the meaning of an expression is built up from the meaning of its parts. Given that the meaning assigned to a string depends on the meaning assigned to the things it is built up from, any string that can be generated in two ways can be given two meanings:

The thing that terrified him was climbing up the drainpipe

This kind of ambiguity is a real pain for language users, and artificial languages are carefully designed not to exhibit it. If we can organise matters so that each formula has precisely one proof or certificate, and uniquely recoverable from the formula at that, then we are freed from the need to explicitly associate to each formula a proof or certificate, and we are thereby allowed to think of a formula simply as a formula.

In chapters 4 and 5 we will use recursion on the engendering relations to assign meaning to formulæ of two families of languages called **propositional** and **predicate** languages. For the moment we will merely set up the syntax of these languages and leave the recursive definition of semantics until later.

2.2.1 Propositional languages

An alphabet of propositional logic contains infinitely many **variables**, also known as **propositional letters**, also known as **literals**; and **connectives** such as $\wedge$, $\vee$, $\rightarrow$, $\longleftrightarrow$, NAND and NOR; and finally bits of punctuation like '(' and ')'.

To be specific, let us say that a propositional letter is one of the letters 'p', 'q' or 'r' with primes attached (so that we have infinitely many of them). Miniexercise: Verify that this makes the set of propositional letters into a regular language over the alphabet { 'p', 'q', 'r' '' }

A propositional language is a set of formulæ over this alphabet recursively generated as follows:

(i) a propositional letter is a formula;

(ii) if p and q are formulæ, so are $(p \vee q)$, $(p \rightarrow q)$, and so on.

If P is a propositional alphabet, the propositional language over P will be written $\mathcal{L}(P)$.

2.2.2 Predicate languages

A predicate language is the richer kind of thing that contains formulæ like $(\forall x)(\forall y)(\forall z)(R(x,y) \wedge R(y,z) \rightarrow R(x,z))$. To be rigorous about it we would have to say something like the following:

(i) A quantifier is $\forall$ or $\exists$.

(ii) A variable is one of the letters 'x', 'y' or 'z' with a number of primes appended to it.

(iii) A predicate letter is an uppercase letter of the Roman alphabet. Recall from page 9 the function called **arity** that takes a predicate letter to the number of arguments it is supposed to have.[1]

(iv) A function letter is a lowercase letter of the Roman alphabet other than 'x', 'y' or 'z'. Function letters have arities the way predicate letters do.

(v) An atomic formula is a predicate letter followed by the appropriate number (the **arity** of that predicate letter) of terms all enclosed within a pair of parentheses and demarcated by commas – for example, $F(f(x), y, g(z))$.

---

[1] On page 9 arity was a quantity associated with a function rather than with a piece of syntax potentially denoting that function. This is an example of use-mention confusion. See the White Knight's song in *Through the Looking glass and what Alice found there*.

(vi) A term is either (i) a variable or (ii) a function letter followed by the appropriate number of terms enclosed in a set of parentheses and demarcated by commas, as it might be '$f(g(x), y)$'.

(vii) A molecular formula is either (i) an atomic formula, or (ii) a boolean combination of molecular formulæ or (iii) the result of hanging a quantifier-with-a-variable in front of a molecular formula.

A function letter might have arity 0, in which case it is a constant symbol. A predicate letter might have arity 0, in which case it is a propositional letter;

A **negatomic** formula is the negation of an atomic formula.

A quantifier $\forall x$ or $\exists x$ always comes equipped with brackets; thus '$(\forall x)(\ldots)$'. The material between the second left and the second right bracket is said to be **within the scope** of the quantifier. If (as in this case) the variable after the quantifier is 'x', then every occurrence of 'x' within the scope is **bound**. An occurrence that is not bound is **free**. Naturally we have the same idea of free and bound variables in lambda calculus too.

The quantifier '$(\exists ! x)\ldots$' is to be read: "there is a unique x such that $\ldots$". If a is a thing that is ϕ, then it is a **witness** to the formula '$(\exists x)\phi(x)$'.

One particularly simple example of a predicate language we will need later is the language of set theory. We characterise it formally by saying that it is a language we would say *in* or *of* predicate calculus with equality and one primitive binary predicate letter '$\in$'. This information is laid down in the **signature**. For example, the signature of set theory is equality plus one binary predicate; the signature of the language of first-order Peano arithmetic has slots for one unary function symbol, one nullary function symbol (or constant) and equality. A signature is something even more abstract than a set of predicate letters and function letters. It is what remains after we throw away the symbols but remember how many of each variety we have. Cricket and baseball have the same signature. Well, more or less! They can be described by giving different values to the same set of parameters. Rings and integral domains have the same signature.

When your mail-order kitset arrives, with the pieces and instructions for how to build your own algebraicly closed field, somewhere buried in the polystyrene chips you have a piece of paper (the "manifest") that is what the signature is in this example. It tells you how many objects you have of each kind, but it does not tell you what to do with them. Instructions on what you do with the objects come with the axioms (instructions for assembly).

2.2.3 Intersection-closed properties and Horn formulæ

A rectype is the intersection of all sets containing certain founders and closed under certain constructors. The property of containing certain founders and being closed under certain constructors – call it **closed** for the moment – has the feature that the intersection of a family of closed sets is also closed. So the intersection of all of them is also closed.

A property that is preserved under intersection in this way is said to be **intersection-closed**: a property of sets is intersection-closed iff the intersection of any number of sets having that property also has that property. Intersection-closed properties give rise to a notion of closure: if X is a set that lacks some intersection-closed property F, then the intersection of all supersets of X that do have F is itself F, is the least superset of X that is F and it is commonly called the F-closure of X. Standard examples include the convex hull of a set of points in a vector space and transitive closure of relations.

At first blush you might think that this is slightly more general than declaring a rectype. Interestingly, this is not so. It turns out that any intersection-closed property $F(X)$ can be twisted into a form where it says that X is closed under certain functions.

Let us illustrate this by recalling the intersection-closed properties of transitivity and symmetry from page 10:

$$(\forall xyz)(R(x,y) \land R(y,z) \to R(x,z))$$

$$(\forall xy)(R(x,y) \to R(y,x)).$$

We can see that a relation is transitive iff it is closed under the function that accepts $\langle x, y \rangle$ and $\langle y, z \rangle$ and returns $\langle x, z \rangle$. It is symmetric iff it is closed under the function $\lambda x.\langle \mathtt{snd}(x), \mathtt{fst}(x) \rangle$ that flips ordered pairs around.

Notice now that these two definitions have a syntactic peculiarity: the stuff inside the quantifiers is of the form

$$(\bigwedge_{i \in I} p_i) \to q,$$

where the p_i and q are atomic (not even negatomic – just atomic). I

may be empty, and q may be $\bot$. '$\bot$' is the constant symbol constrained
to evaluate always to `false`: this matches its other use as the symbol
denoting the bottom element of a poset. Formulæ like this are called
horn clauses. We will say that a property F of relations is **captured
by horn clauses** if the assertion that R has property F is expressed
by a formula that is a list of universal quantifiers enclosing a body that
is a Horn clause whose atomic parts are fragments like '$R(x, y)$', which
just glue together with R's the variables mentioned in the universal
quantifiers. The graph of a relation with a Horn property can always be
thought of as a set of ordered tuples closed under some function.

To put it roughly:

REMARK 7 *The following are equivalent for a property F:*
F is intersection-closed.
F is captured by Horn clauses.
The extension (graph) of F is a rectype.

Horn clauses are the syntactic manifestation of rectypes and have
computational significance in various ways. For example, because there
is a single conclusion there is precisely one pair to add at each step,
and we know which it is so the construction is *deterministic*. This is in
contrast to the process of adding ordered pairs to a relation to get one
satisfying connexity: $R(x, y) \lor R(y, x)$. If neither $\langle x, y \rangle$ nor $\langle y, x \rangle$ are
present, which do we add? A further attractive feature of the process
of adding ordered pairs in a way dictated by a Horn clause is that it
is insensitive to the order in which we scan the ordered pairs we have
already. Contrast this with the construction of an invitation list for a
party, which (for these purposes) is a gathering of people all of whom are
on speaking terms with all of the others. If I draw up a list of invitations
by inviting everyone who is on speaking terms with those already on the
list so far, the final list will be very sensitive to the order in which I store
the names of my friends in my address book.

Horn properties and rectypes have other nice features. A convex set in
a vector space can be thought of as a rectype, and it illustrates graphically how well Horn properties behave in relation to projection and
directed unions (see page 15). The projection of a convex set in a vector
space is likewise a convex set: the shadow cast by a convex solid figure
is a convex plane figure. The set of convex subsets of E^n is closed under
directed unions.

2.2.4 Do all well-founded structures arise from rectypes?

All rectypes give rise to well-founded relations: do all well-founded relations arise from rectypes? For most practical purposes the answer appears to be 'yes'. Counterexamples would be interesting, but no one has ever fomulated the question precisely enough for us to know what we would be looking for. What do we mean by 'arise', exactly?

3

Partially ordered sets

3.1 Lattice fixed point theorems

A **fixed point** for a function f is an argument x such that $f(x) = x$. This is an important concept because many useful mathematical facts can be expressed by assertions that say that certain functions have fixed points. For example, the equation $p(x) = 0$ has a solution iff the function $\lambda x.(p(x) - x)$ has a fixed point. This gives us a motive to seek methods for showing that functions have fixed points: fixed point theorems are useful in the search for solutions to equations.

3.1.1 The Tarski-Knaster theorem

THEOREM 8 *(The Tarski-Knaster theorem) Let $\langle X, \leq \rangle$ be a complete lattice and f an order-preserving map $\langle X, \leq \rangle \to \langle X, \leq \rangle$. Then f has a fixed point.*

Proof: Set $A = \{x : f(x) \leq x\}$ and $a = \bigwedge A$. (A is nonempty because it must contain $\bigvee X$.) Since f is order-preserving, we can say that if $f(x) \leq x$, then $f^2(x) \leq f(x)$, and so $f(a)$ is also a lower bound for A as follows. If $x \in A$, we have $f(x) \leq x$, whence $f^2(x) \leq f(x)$, so $f(x) \in A$ and $a \leq f(x)$. But $f(x) \leq x$ so $f(a) \leq x$ as desired. But a was the *greatest* lower bound, so $f(a) \leq a$ and $a \in A$. But then $f(a) \in a$ since $f``A \subseteq A$, and $f(a) \geq a$ since a is the greatest lower bound. ∎

   This proof of theorem 8 shows not only that increasing functions have fixed points but that they have *least* fixed points. This gives us the existence of inductively defined sets because the operation of taking a set and adding to it the result of applying all the constructors once to all its members is increasing (with respect to $\subseteq$). The above definition of the element a echoes precisely the declaration of $\mathbb{N}$ as an intersection of a family of sets. Compare $\bigwedge \{x : f(x) \leq x\}$ with $\bigcap \{X : (S\text{``}X \cup \{0\}) \subseteq X\}$.

EXERCISE **16** *Prove that every monotone function on a complete lattice has a greatest fixed point.*

A least fixed point (think $\mathbb{N}$) has an induction principle, and a greatest fixed point has a co-induction principle. What might this co-induction principle be? Something will belong to a coinductive datatype as long as there is no good finite reason for it not to.

THEOREM **9** *(Schröder-Bernstein) If f is an injection from A into B and g and injection from B into A, then there is a bijection between A and B.*

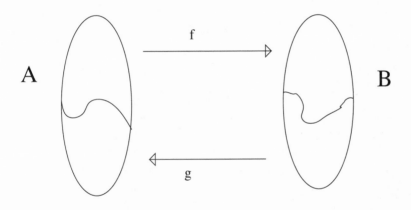

Fig. 3.1. The Schröder-Bernstein theorem

The function $\lambda X.(A \setminus g\text{``}(B \setminus f\text{``}X))$ is a monotone map from $\mathcal{P}(A)$ into itself. $\lambda X.f\text{``}X$ is monotone; complementation in $B$ is antimonotone; $\lambda Y.g\text{``}Y$ is monotone and complementation in A is antimonotone. the composition of two antimonotone functions is antimonotone, so the function $\lambda X.(A \setminus g\text{``}(B \setminus f\text{``}X))$ is a monotone map from $\mathcal{P}(A)$ into itself.

If now X is a fixed point for $\lambda X.(A \setminus g\text{``}(B \setminus f\text{``}X))$, we find that $f|X \cup (g^{-1}|(A \setminus X)$ is a bijection between A and B. ∎

Further applications include the existence of transitive closures of relations. Consider the complete lattice $\mathcal{P}(X \times X)$ and let f be the function $\lambda R.R \cup R^2$. Any fixed point for this function is a transitive relation. If

N is a binary relation on X, then the least fixed point of $\lambda R.R \cup R^2$ that is above N is the transitive closure of N. Is there a fixed point above N? Yes, because the the upper set of points above any given point in a complete lattice is also a complete lattice and we can use theorem 8 again.

3.1.2 Witt's theorem

We say $f : X \to X$ is **inflationary** if $(\forall x \in X)(x \leq f(x))$.

THEOREM 10 *Every inflationary function from a chain-complete poset into itself has arbitrarily late fixed points.*

Proof: Let $\langle X, \leq \rangle$ be a chain-complete poset, f an inflationary function $X \to X$ and x a member of X. We will show that f has a fixed point above x.

The key device is the inductively defined set of things obtainable from x by repeatedly applying f and taking sups of chains – the smallest subset of X containing x and closed under f and sups of chains. Let us call this set $C(x)$. Our weapon will be induction.

We will show that $C(x)$ is always a chain. Since it is closed under sups of chains, it must therefore have a top element and that element will be a fixed point.

Let us say $y \in C(x)$ is **normal** if $(\forall z \in C(x))(z < y \to f(z) \leq y)$. We prove by induction that if y is normal, then $(\forall z \in C(x))(z \leq y \vee f(y) \leq z)$. That is to say, we show that – for all normal y – $\{z \in C(x) : z \leq y \vee f(y) \leq z\}$ contains x and is closed under f and sups of chains and is therefore a superset of $C(x)$. Let us deal with each of these in turn.

(i) (Contains x) $x \in \{z \in C(x) : z \leq y \vee f(y) \leq z\}$ because $x \leq y$. ($x \leq y$ because x is the smallest thing in $C(x)$ – by induction!) The set of things $\geq x$ contains x, is closed under f and sups of chains and is therefore a superset of $C(x)$.

(ii) (Closed under f) If $z \in \{z \in C(x) : z \leq y \vee f(y) \leq z\}$, then either

   (a) $z < y$, in which case $f(z) \leq y$ by normality of y and $f(z) \in \{z \in C(x) : z \leq y \vee f(y) \leq z\}$; or

   (b) $z = y$, in which case $f(y) \leq f(z)$ so $f(z) \in \{z \in C(x) : z \leq y \vee f(y) \leq z\}$; or

   (c) $f(y) \leq z$, in which case $f(y) \leq f(z)$ (f is inflationary) and $f(z) \in \{z \in C(x) : z \leq y \vee f(y) \leq z\}$.

(iii) (Closed under sups of chains) Let $S \subseteq \{z \in C(x) : z \leq y \vee f(y) \leq z\}$ be a chain. If $(\forall z \in S)(z \leq y)$, then $sup(S) \leq y$. On the other hand, if there is $z \in S$ s.t. $z \nleq y$, we have $f(y) \leq z$ (by normality of y); so $sup(S) \geq f(y)$ and $sup(S) \in \{z \in C(x) : z \leq y \vee f(y) \leq z\}$.

Next we show that everything in $C(x)$ is normal. Naturally we do this by induction: the set of normal elements of $C(x)$ will contain x and be closed under f and sups of chains.

(i) (Contains x) Vacuously!

(ii) (Closed under f) Suppose $y \in \{w \in C(x) : (\forall z \in C(x))(z < w \rightarrow f(z) \leq w\}$. We will show $(\forall z \in C(x))(z < f(y) \rightarrow f(z) \leq f(y))$. So assume $z < f(y)$. This gives $z \leq y$ by normality of y. If $z = y$, we certainly have $f(z) \leq f(y)$, as desired, and if $z < y$, we have $f(z) \leq y \leq f(y)$.

(iii) (Closed under sups of chains) Suppose $S \subseteq \{w \in C(x) : (\forall z \in C(x))(z < w \rightarrow f(z) \leq w)\}$ is a chain. If $z < sup(S)$, we cannot have $(\forall w \in S)(z \geq f(w))$ for otherwise $(\forall w \in S)(z \geq w)$ (by transitivity and inflationarity of f), so for at least one $w \in S$ we have $z \leq w$. If $z < w$, we have $f(z) \leq w \leq sup(S)$ since w is normal. If $z = w$, then w is not the greatest element of S, so in S there is $w' > w$ and then $f(z) \leq w' \leq sup(S)$ by normality of w'.

If y and z are two things in $C(x)$, we have $z \leq y \vee f(y) \leq z$ by normality of y, so the second disjunct implies $y \leq z$, whence $z \leq y \vee y \leq z$. So $C(x)$ is a chain as promised, and its sup is the fixed point above x whose coming was foretold. ∎

3.1.3 Exercises on fixed points

EXERCISE 17 *Show that the fixed point of theorem 8 is $\leq_X$-minimal.*

EXERCISE 18 *Let $\langle A, \leq \rangle$ and $\langle B, \leq \rangle$ be total orderings with $\langle A, \leq \rangle$ isomorphic to an initial segment of $\langle B, \leq \rangle$ and $\langle B, \leq \rangle$ isomorphic to a terminal segment of $\langle A, \leq \rangle$. Show that $\langle A, \leq \rangle$ and $\langle B, \leq \rangle$ are isomorphic.*

EXERCISE 19 *(The Gale-Stewart theorem) Return to the generalised game described on p. 36. This time, if the game goes on forever,* II *wins.*

Provide a formal notion of **winning strategy** *for games of this sort. Use Witt's theorem to prove that one or the other player must have a winning strategy in your sense.*

EXERCISE 20 *What might the well-founded part of a binary relation be? Use one of the fixed point theorems to show that your definition is legitimate.*

EXERCISE 21 *An examination question*

(i) State and prove the Tarski-Knaster fixed point theorem for complete lattices.

(ii) Let X and Y be sets and $f : X \to Y$ and $g : Y \to X$ be injections. By considering $F : \mathcal{P}(X) \to \mathcal{P}(X)$ defined by

$$F(A) = X \setminus g``(Y \setminus f``X)$$

or otherwise, show that there is a bijection $h : X \to Y$.

Suppose U is a set equipped with a group Σ of permutations. We say that a map $s : X \to Y$ is piecewise–Σ just when there is a finite partition $X = X_1 \cup \ldots \cup X_n$ and $\sigma_1 \ldots \sigma_n \in \Sigma$, so that $s(x) = \sigma_i(x)$ for $x \in X_i$. Let X and Y be subsets of U, and $f : X \to Y$ and $g : Y \to X$ be piecewise–Σ injections. Show that there is a piecewise–Σ bijection $h : X \to Y$.

If $\langle P, \leq_P \rangle$ and $\langle Q, \leq_Q \rangle$ are two posets with order-preserving injections $f : P \to Q$ and $g : Q \to P$, must there be an isomorphism? Prove or give a counterexample.

3.2 Continuity

Notice that neither theorem 8 nor theorem 10 make any assumptions about the continuity of the functions they produce fixed points for. To appreciate the significance of this point, attempt the following exercise.

EXERCISE 22 *Let $\langle X, \leq_X \rangle$ be a complete partial order and f a monotone function $\langle X, \leq_X \rangle \to \langle X, \leq_X \rangle$. Show that $\{x : x = f(x)\}$ is a complete lattice.*

One naturally spots immediately that any set F of fixed points for f has a sup in X. Equally naturally one expects next to be able to prove that this fixed point is itself fixed. Why should one expect this? There are two reasons, both of which bear examination.

(i) One might expect that the subposet of fixed points for f inherits not only the ordering from $\langle X, \leq_X \rangle$ but also the $\bigvee$ and $\bigwedge$ as well. That is to say, one expects the subposet of fixed points to be not only a subposet but a sub complete lattice as well. This is a natural thing to expect because, in most cases where one has two algebras and the carrier set of the second is a subset of the carrier set of the first, the operations of the second are restrictions of the operations of the first: subgroups of groups have the same multiplication as the group of which they are subgroups; multiplication of rationals is a restriction of multiplication of the reals, and so on. It is unusual to encounter a widget in circumstances where a naturally presented subset of its carrier set has a widget structure but is not a substructure of the original widget.

(ii) There are various natural concepts of continuity that one might be unconsciously invoking, and they will ensure that the sup of X in the poset of fixed points is the same as its sup in $\langle X, \leq_X \rangle$. These ideas of continuity bear examination in turn.

The roots of all ideas of continuity lie in the real line. Topology arose from an endeavour to develop for use elsewhere the notion of a continuous function from the reals to the reals. It is a powerful development because it remains useful even when the domain and range of the putatively continuous functions lack order structure. In the present context we still have order structure, and we can develop the same original ideas in a different direction.

Let $\langle X, \leq_X \rangle$ be a complete lattice. $f : X \to X$ can be made to act on $\mathcal{P}(X)$ in two ways. Given $Y \subseteq X$, one can take the sup and then apply f, or one can apply f setwise to Y and take the sup of the values. Are the two results the same? If they are, we say f is continuous. Notice that if X is $\Re$, then this agrees with the usual definition of a continuous function $\Re \to \Re$, at least for nondecreasing functions.

So $f : X \to X$ is **continuous** if $(\forall X' \subseteq X)(\bigvee(f``X') = f(\bigvee(X')))$, that is to say, if the following diagram commutes.

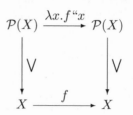

If we write X^α for the set of subsets of X that are ranges of increasing[1] X-valued sequences of length α, we then say that f is α-**continuous** if the next diagram commutes:

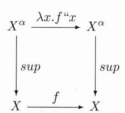

We are getting slightly ahead of ourselves here, as we have not met ordinals yet and will not until chapter 7. However, for the reader to make sense of the discussion here it will suffice for them to think of ordinals as numbers that measure the lengths of sequences. Of course you have not ever had to worry about α-continuity for any α other than ω because, as you know, whenever a is a least upper bound of a set of reals X there is an increasing sequence x_0, x_1 ... indexed by the naturals whose limit is a (see exercise 6 page 28). So the only kind of continuity of functions $\Re \to \Re$ that matters is ω-continuity, where ω is the order type of the naturals in their natural order. So you would not have learned a general concept of α-continuity from the reals.

In algebra one has functions like $\lambda A.\{ab : a, b \in A\}$ that take a set of group elements to another set of group elements. This example too is ω-continuous. It seems that it is ω-continuous because it has finite character. However, even some functions whose character is less obviously finite are ω-continuous. Consider the function that sends a set to the set of all its finite subsets. Even this is ω-continuous: if $a_1 \subseteq a_2 \subseteq a_3 \subseteq \ldots$ and x is a finite subset of $a_1 \cup a_2 \cup a_3 \cup \ldots$, then it can only meet finitely

---

[1] We need this condition because without it any β-sequence with $\beta < \alpha$ could be padded out to an α-sequence, with the effect that α-continuity of a function f would imply β-continuity for all $\beta < \alpha$.

many $a_{i+1} \setminus a_i$, and so it is already a subset of some a_i. So $\lambda x.$(set of finite subsets of x) is ω-continuous. In contrast, $\lambda x.$(set of countable subsets of x) is not! Think about how to apply this reasoning to the set of countable subsets of a set. If you know about ordinals already, you may ask yourself: what can we deduce about an ordinal α if we are told that the function $\lambda x.$(set of countable subsets of x) is α-continuous? We will return to this in chapter 8. Finally, $\mathcal{P}$ is an example of a function that is monotone but not α-continuous for any α.

Given that many natural functions are continuous in one sense or another, it is natural to wonder if one can weaken the requirement that the domain and range should be a continuous lattice if the only functions for which one seeks fixed points are continuous. An intuition that is very appealing in this context is the idea of iterating a continuous function and looking at the limit of the points obtained. Is the sup of $\{x, f(x), f^2(x) \ldots f^n(x) \ldots\}$ a fixed point for f, if f is continuous? It will be – as long as it exists! What condition can we put on the lattice that will ensure that this limit exists? Well, if f is monotone increasing, then the set $\{f^n(x) : x \in \mathbb{N}\}$ will be a chain, so all that is necessary is to suppose that the lattice has sups of all chains of length ω, which is a weaker condition than the existence of sups of all subsets.

This is susceptible of refinements that we will not pursue here: arguments like this will enable us to show that if a poset has sups of all chains of length α, then if f is α-continuous, then f will have a fixed point.

In fact, not only do we not need the domain and range to be a complete lattice, we do not even need it to be a lattice at all. The condition on existence of sups of chains that does the business for us does not imply the existence of sups of two incomparable elements. Our next example illustrates this.

Let us write "Z $\rightharpoonup$ Z" for the set of partial maps from the integers into itself. (The funny arrow is not defective – it really is meant to have only one fletch, and its LaTeX symbol is \rightharpoondown. $X \rightharpoonup Y$ is the set of partial functions from X to Y.)

Identify the maps with their graphs and partially order Z $\rightharpoonup$ Z by set inclusion. This makes it a chain-complete poset under $\subseteq$. It inherits its structure from the complete lattice $\langle \mathcal{P}(\text{Z} \times \text{Z}), \subseteq \rangle$ – of which it is a substructure in terms of the jargon on page 7. Now consider the map

metafact: $\lambda f.\lambda n.\text{if } n = 0 \text{ then } 1 \text{ else } n * f(n-1)$

Notice that `metafact` is ω-continuous, and that the poset of partial maps $Z \rightharpoonup Z$ partially ordered by inclusion has sups of ω-chains. So `metafact` will have a fixed point.

Check for yourself that any fixed point for this satisfies the recursion that characterises the factorial function. In fact there are many fixed points for `metafact`, but the one we are after is the *least* one, which we can obtain by iteration in the obvious way. The least fixed point is the only fixed point that contains no information beyond that obtainable from the recursion. The recursion only tells us what to do to natural numbers, so the least fixed point is undefined everywhere else.

This illustrates how the extra generality (of chain-complete posets over complete lattices) matters. The set of partial maps $Z \rightharpoonup Z$ partially ordered by inclusion is an important object, but it is not a complete lattice (two functions which disagree on even one argument have no common upper bound), so we cannot use the Tarski-Knaster theorem to show that things like `metafact` have fixed points.

3.2.1 Exercises on lattices and posets

EXERCISE 23 *Let $\mathcal{O}(X)$ be the lattice of open sets of a topological space X. Show that it is a complete lattice under inclusion. Is it distributive? There are two infinitary distributive laws: $x \wedge \bigvee A = \bigvee \{x \wedge a : a \in A\}$ and $x \vee \bigwedge A = \bigwedge \{x \vee a : a \in A\}$. Which of these does it satisfy?*
Consider the map

$$F : \mathcal{O}(X) \to \mathcal{O}(X); \quad F(U) = \mathrm{int}(X \setminus \mathrm{int}(X \setminus U)),$$

where $\mathrm{int}(A)$ is the interior of $A \subseteq X$. Show that F is order-preserving. Is F continuous?

EXERCISE 24 *Consider the following functions $F : \mathcal{P}(\mathbb{N}) \to \mathcal{P}(\mathbb{N})$.*

(i) $F(A) = \{1\} \cup \{2n : n \in A\} \cup \{3n : n \in A\}$.
(ii) $F(A) = A \cup \{2n : n \notin A\}$.
(iii) $F(A) = \{2\} \cup \{ab : a, b \in A\}$.
(iv) $F(A) = \{n : (\exists m \in A)(n \leq m)\}$.
(v) $F(A) = \mathbb{N} \setminus A$.

In each case determine whether or not F is ω-continuous. In the cases where F is ω-continuous, identify the least fixed point of f.
In the cases when F is not ω-continuous, determine whether or not F has a fixed point.

EXERCISE 25 *The proof of theorem 3.1.1 uses the function*

$$F : \mathcal{P}(X) \to \mathcal{P}(X); \quad F(A) = X \setminus g``(Y \setminus f``A),$$

where $f : X \to Y$ *and* $g : Y \to X$ *are injections. Is this function* ω*-continuous?*

3.3 Zorn's lemma

Zorn's lemma is one of a collection of interdeducible assertions.

Zorn's lemma: *Every poset in which every chain has an upper bound has a maximal element.*

Do not worry too much about whether or not it might be true. We can use it to prove various convenient generalities, like:

(i) Every vector space has a basis.

(ii) Every set can be well-ordered.

(iii) Given any two sets, there is a injection from one into the other.

(iv) The Axiom of Choice: If X is a set of nonempty sets, there is a function $f : X \to \bigcup X$ s.t. $(\forall x \in X)(f(x) \in x)$.

(v) Countable choice: Like (iv), but X is required to be countable.

(vi) Every surjection has a right inverse.

(vii) Tikhonov's theorem: A product of compact spaces is compact.

(viii) The Jordan-König theorem (see page 196).

(ix) (Nilson-Schreier) Every subgroup of a free group is free.

(x) Every connected graph has a spanning tree.

(xi) In a chain-complete poset, every directed subset has a sup.

EXERCISE 26 *Deduce items* (i), (ii), (iii), (iv), (vi), (x) *and* (xi) *from Zorn's lemma.*

Most of items (i) to (xi) above are not only implied by Zorn's lemma but are actually equivalent to it. Item (v) is known to be weaker than Zorn's lemma, and it is not known whether (ix) implies Zorn's lemma or not. All the others are known to, but the proofs in the other direction are beyond the scope of this book. Theorem 10 was proved specifically to deduce Zorn's lemma from the axiom of choice.[1]

---

[1] Deducing Zorn from AC was a well-known fiddly task and in some ways remains so. I learned the idea of using Witt's theorem for it from my colleague, Peter Johnstone.

EXERCISE **27** *Use Witt's theorem and the axiom of choice to prove that every chain-complete poset has a maximal element.*

Show that, for any poset $\langle X, \leq_X \rangle$, the collection of chains in it, partially ordered by $\subseteq$, is a chain-complete poset.

Then deduce Zorn's lemma from this last assertion.

(You may need the hint that this generalises the construction of the reals as the completion of the rationals.)

EXERCISE **28** *Prove the following implications:*

(vii) $\rightarrow$ (iv);

(ii) $\rightarrow$ (iv);

(x) $\rightarrow$ (iv).

3.3.1 Exercises on Zorn's lemma

EXERCISE **29** *Recall that one partial order $\leq_2$ on a set P extends the partial order $\leq_1$ just when $a \leq_1 b$ implies $a \leq_2 b$ for all a, b in P. Use Zorn's lemma to show that every partial order can be extended to a total order. This is the* **order extension principle**.

EXERCISE **30** *Use Zorn's lemma to prove that every subspace $U \leq V$ of a vector space has a complementary subspace (that is, there is $W \leq V$ with $V = U \oplus W$).*

3.4 Boolean algebras

Recall that a boolean algebra is a distributive complemented lattice.

Notice that all the axioms for boolean algebras are Horn. So we can talk about the boolean algebra generated by a set of elements, and we can talk about products and substructures of boolean algebras. Indeed, they are a special kind of Horn formula, being of the form $(\forall x_1 \ldots x_n)$ hung on the front of a conjunction of a lot of equations. Theories axiomatised by universal closures of conjunctions of equations are said to be **algebraic**. (Look ahead to page 95 if you do not understand this and are in a hurry to find out.)

The most natural examples of boolean algebras are power set algebras: the set of all subsets of a given set, partially ordered by set inclusion, with union, intersection and complement.

3.4.1 Filters

A filter in a boolean algebra is a subset F of the carrier set that is closed under $\geq$ and $\wedge$, that is, it satisfies the two conditions:

$$x \in F \wedge x \leq y \rightarrow y \in F$$

and

$$x, y \in F \rightarrow x \wedge y \in F.$$

Notice that these conditions are Horn, so an intersection of filters is a filter and a directed (page 15) limit of filters is a filter. The fact that it is Horn also means we can talk about the filter generated by a set, which is of course the smallest filter that is a superset of the set given. A filter in the power set algebra $\langle \mathcal{P}(X), \subseteq \rangle$ is said to be a filter **on** X. We should think of a filter on X as a concept of largeness (of subsets of X). This seems reasonable if we reflect on the easiest examples: the cofinite subsets of $\mathbb{N}$ (these are the sets X such that $\mathbb{N} \setminus X$ is finite) are clearly large in some sense. This motivates two other clauses that we almost always assume and which one can easily overlook.

(i) Proper filters. According to the definition of filter, the whole algebra is a filter. However, it is not a **proper** filter. All other filters are proper. If the filter generated by a set of elements is proper, we say the set is a **filter base**.

(ii) Nonprincipal filters. There are pathological filters that do not accommodate the "largeness" intuition. If b is any element of a boolean algebra $\mathcal{B}$, then $\{b' \in \mathcal{B} : b' \geq b\}$ is a filter in $\mathcal{B}$. It is the **principal filter generated by** b. We will think of principal filters as pathological and will not be interested in them. The remaining filters are **nonprincipal**, like the filter of cofinite subsets of $\mathbb{N}$ we saw earlier.

EXERCISE 31 *Check that the filters in a fixed boolean algebra form a complete poset and that the proper filters form a chain-complete poset.*

Let F be a filter in a boolean algebra $\mathcal{B}$. $\{\neg y : y \in F\}$ is an **ideal**. Indeed it is the **dual ideal** to F. Ideals in boolean algebras are so-called because they correspond to ideals in boolean rings. Boolean rings?

EXERCISE 32 *A **boolean ring** is a ring with a 1 such that for all elements x, $x^2 = x$.*

(i) *Describe operators and equations that show that the theory of boolean rings is an algebraic theory.*

(ii) *Show that a boolean ring satisfies the equations $x + x = 0$ and $xy = yx$. Deduce (by considering additive groups) that every finite boolean ring has order a power of 2.*

(iii) *Show that a boolean algebra becomes a boolean ring with multiplication given by $\wedge$ and addition defined by $x + y = (x \wedge \neg y) \vee (y \wedge \neg x)$.*

(iv) *Conversely, find definitions of 0, 1, $\vee$, $\wedge$ and $\neg$ in terms of $+$, 0, 1 and so on so that a boolean ring becomes a boolean algebra.*

(v) *Which boolean rings are integral domains?*

DEFINITION 11 *A filter F satisfying any of the conditions below is said to be an* **ultrafilter**.

(i) *F is $\subseteq$-maximal among proper filters.*

(ii) *$(\forall x \in \mathcal{B})(x \in F \vee \neg x \in F)$.*

(iii) *For all a, $b \in \mathcal{B}$, if $(a \vee b) \in F$, then either $a \in F$ or $b \in F$. (F is* **prime**.*)*

The ideal dual to an ultrafilter is **prime**.

EXERCISE 33 *Prove that the definitions of definition 11 are equivalent.*

There are natural examples of filters on sets: we saw earlier the filter of cofinite subsets of $\mathbb{N}$, and indeed for any infinite set X the collection of cofinite subsets of X is a filter on X. Unfortunately, the only natural examples of ultrafilters are trivial. If x is any element of a set X, then $\{X' \subseteq X : x \in X'\}$ is a principal ultrafilter on X, and, unless we assume something like the axiom of choice, this is the only kind of ultrafilter whose existence can be demonstrated.

We tend to use $\mathcal{CALLIGRAPHIC}$ font capitals for variables ranging over ultrafilters.

If we do assume the axiom of choice we can prove that there are ultrafilters aplenty:

THEOREM 12 *(The prime ideal theorem) Every boolean algebra has an ultrafilter.*

Proof: Consider the set of filters in a boolean algebra $\mathcal{B}$. They are partially ordered (by $\subseteq$, as we have remarked earlier in exercise 31); also, any $\subseteq$-chain of filters has an upper bound (which is simply the union of them all) so the assumptions of Zorn's lemma are satisfied. Therefore there are maximal filters. These are ultra, by exercise 33. ∎

Since proving that there are ultrafilters is the same as proving that there are maximal ("prime") ideals, the name should not cause puzzlement: it is simply a question of which terms you propose to think in.

Since, as we noted on page 15, upper sets in (complete) posets are (complete) posets, we can even prove the apparently stronger assertion that every filter in a boolean algebra $\mathcal{B}$ can be extended to an ultrafilter. It is not in fact any stronger because, if we seek an ultrafilter extending a given filter F, we form the quotient algebra $\mathcal{B}/F$, use theorem 12 to find an ultrafilter, and then form the set of all elements of $\mathcal{B}$ that got sent to the ultrafilter in the quotient. This set is an ultrafilter extending F.

In fact, by being careful in the choice of a chain-complete poset we can even prove:

EXERCISE 34 *If $\mathcal{B}$ is a boolean algebra with nonprincipal filters, then it has a nonprincipal ultrafilter.*

Algebras have products and quotients. A homomorphism from $\mathcal{A}$ to $\mathcal{B}$ is a map h such that, if a tuple $\vec{a}$ of elements of $\mathcal{A}$ stands in some (atomic) relation R in $\mathcal{A}$, then the tuple $h(\vec{a})$ stands in the same relation R in $\mathcal{B}$. In fact, there is usually more one can say about homomorphisms than this. In the case of boolean algebras (which are the only algebras we are going to be interested in here), any filter gives rise to a homomorphism. As noted earlier, a filter corresponds to a notion of largeness. Thus, if we have a filter F in a boolean algebra $\mathcal{B}$, it is natural to think of b and b' in $\mathcal{B}$ being similar if their symmetric difference $b \Delta b'$ is *small*, which is to say, its complement is in the filter. Thus we have $b \sim_F b'$ iff the complement of $(b \Delta b') \in F$.

EXERCISE 35

(i) *Check that $\sim_F$ is the same as $(\exists c \in F)(c \wedge b = c \wedge b')$.*

(ii) *Check that $\sim_F$ is a congruence relation for the boolean operations.*

(iii) *Prove that the function sending elements of $\mathcal{B}$ to their equivalence classes is a boolean algebra homomorphism.*

DEFINITION 13 *The algebra whose elements are equivalence classes* *under* $\sim_F$ *is the* **quotient algebra modulo** F. *The* **kernel** *of a homomorphism of boolean algebras is the set of elements sent to 0.*

This enables us to prove the *Stone representation theorem.* A representation theorem you already know is the representation theorem for groups: every group is (isomorphic to) a group of permutations of a set. The most obvious examples of boolean algebras all have sets as their elements and set inclusion ($x \subseteq y$) as their partial order, but not all do: quotient algebras typically do not. The Stone representation theorem is the assertion that nevertheless

THEOREM 14 *(Stone's representation theorem) Every boolean algebra is isomorphic to a boolean algebra whose elements are sets, whose partial order is* $\subseteq$, *and whose* $\vee$ *and* $\wedge$ *are* $\cup$ *and* $\cap$.

Proof: The hard part is to find the isomorphic algebra; the rest is easy. Given $\mathcal{B}$, construct $\mathcal{B}'$ as follows. Send each $b \in \mathcal{B}$ to $\{\mathcal{U} : b \in \mathcal{U}\}$ (the set of all ultrafilters in $\mathcal{B}$ containing b). $\mathcal{B}'$ will be the image of $\mathcal{B}$ in this map. Obviously, if $b \leq c$, then any ultrafilter containing b will contain c but not vice versa, unless $c \leq b$. If b is strictly below c, then consider the principal filter generated by $c \wedge \neg b$. Extend this to an ultrafilter by theorem 12. This ultrafilter will contain c but not b. Thus $b \leq c \longleftrightarrow \{\mathcal{U} : b \in \mathcal{U}\} \subseteq \{\mathcal{U} : c \in \mathcal{U}\}$.

∎

Theorems 12 and 14 are in fact equivalent. Although we used Zorn to prove them, there is no converse. Nevertheless, there is a list of natural assertions equivalent to them, though it is not as long as the list of equivalents of AC. The most interesting item is probably: a product of compact Hausdorff spaces is compact Hausdorff, but that is hard! (See Johnstone 1982)

3.4.2 Atomic and atomless boolean algebras

An **atom** in a boolean algebra is a minimal nonzero element. A boolean algebra is **atomic** if every nonzero element is above an atom. Power set algebras are of course atomic – the atoms are the singletons.

A rich source of atomless boolean algebras are things of the form $RO(\mathcal{T})$, the **algebra of regular open sets**[1] of a topological space $\mathcal{T}$.

---

[1] An open set is regular open if it is the interior of its closure.

The poset of open sets is a Heyting algebra. Atomless boolean algebras will reappear in section 5.6.

3.5 Antimonotonic functions

A function f from a poset into itself is **antimonotonic** iff $(\forall x, y)(x \leq y \to f(y) \leq f(x))$.

Theorem 8 tells us nothing about fixed points for antimonotonic functions, but sometimes one can get results by doing clever ad hoc things. A fact that is sometimes useful is that the composition of two antimonotonic functions is monotonic, and every fixed point for f is also a fixed point for f^2. Look also at question 3.1.3.24.

Fixed points for antimonotonic functions – or at least things with that kind of flavour – crop up inconveniently all over the place. Natural examples in mathematics include finding roots of polynomials. After all, $x^2 - 2 = 0$ has a solution iff the antimonotonic function $\lambda x.(2/x)$ has a fixed point. But in this case there are other techniques we can use, since the poset of reals has extra structure. Complementation in boolean algebras and $\bigcap$ in the poset of all sets are both antimonotonic. Further interesting examples are to be found in linguistics and biology. Two creatures of the same species are supposed to be able to mate and produce viable offspring. This gives rise to a possible definition of a(n extension of a) species as a fixed point for the function

$$\lambda X.\{y : (\forall x \in X)(x \text{ and } y \text{ can be mated to produce viable offspring})\}.$$

(We disregard gender for the moment!) The only trouble is, this function is antimonotonic with respect to $\subseteq$! Let M and F be two sets and $R \subseteq M \times F$. Then the function $m = \lambda X \subseteq M.\{y \in F : (\forall x \in X)(R(x, y))\}$ is an antimonotone function from $\mathcal{P}(M) \to \mathcal{P}(F)$, and similarly $f = \lambda X \subseteq F.\{y \in M : (\forall x \in X)(R(y, x))\}$ is an antimonotone function from $\mathcal{P}(F) \to \mathcal{P}(M)$. Now $f \circ m$ is a monotone map $\mathcal{P}(F) \to \mathcal{P}(F)$ and $m \circ f$ is a monotone map from $\mathcal{P}(M) \to \mathcal{P}(M)$. $f \circ m$ and $m \circ f$ have fixed points by theorem 8. If we take M and F to be the set of (genotypes of) male and female fruit flies, respectively, and $R(x, y)$ to be the binary relation "x and y will produce viable offspring when mated", we find that fixed points for the compositions (either way) of these two maps give rise to things like *species* of fruit flies. Is there any reason to suppose that every member of $M \cup F$ belongs to a fixed point? You may enjoy working out the details.[1]

There are equally important examples from other areas too. In phonetics there is the concept of *allophone*. Two sounds are allophones for a language if the language makes no use of the difference between them. The voiced and unvoiced *th* sounds as in *pith* and *wither* are distinct for native speakers of English in that they can hear that these two sounds are distinct. However, they are equivalent in the sense that there is no pair of English words differing only in that one has a voiced *th* where the other has an unvoiced *th*.[2] (They are not equivalent in this sense in Arabic, for example.) There are other pairs of sounds in English that are indistinguishable in this sense, but they are less striking: front and back 'l's, front and back 'k's, for example. I am not sure about the sounds *sh* and *zh* (as in 'pleasure') for example: I know of no pair of English words that differ only in that one has 'sh' where the other has 'zh'. Let's suppose for the sake of argument that there is no such pair and that these two sounds are indistinguishable in the same sense as the voiced and unvoiced 'th'.

But even if both of these pairs are indistinguishable in that sense, it does not imply that they are as it were *jointly* indistinguishable: there might be two words that differ in that where one has a voiced 'th' and a voiced 'sh' the other has the two unvoiced sounds. What we want is a notion of equivalence of tuples of sounds. That equivalence relation will be a fixed point for an antimonotonic function.

Biology provides us with an important example with the same logical structure: the notion of phenotypic equivalence of two alleles at a locus: if when we swap one for the other it makes no difference to the resulting genotype, we say they are phenotypically equivalent. But if A and a are phenotypically equivalent[3] at one locus and B and b are phenotypically equivalent at another, can we *simultaneously* swap a for A and b for B and *still* not make any difference to the phenotype?

We will revisit these themes briefly in section 4.2.3.

---

[1] It is probably rather important that realistically R^3 is very nearly a subset of R. It is hard to see how to express this or make allowances for it.

[2] Well, very nearly anyway: the only counterexamples to this claim are contrived or obscure: loathe/loth, thy/thigh and thou (as in 'you')/'thou' (as in 'Morrie thou').

[3] There is a convention in the biology literature of using an uppercase letter and the corresponding lowercase letter to denote alleles at a locus, when we are considering only two alleles.

3.6 Exercises

EXERCISE 36

   (i) *Show that $X \subseteq Y \rightarrow \bigcap Y \subseteq \bigcap X$.*
   (ii) *Show that if additionally everything in Y is a subset of something in X, then $\bigcap Y = \bigcap X$.*

EXERCISE 37 *Why do we need ultrafilters in the proof of theorem 14? Why cannot we send b to the set of filters containing b?*

EXERCISE 38 *Let X be an infinite set. Observe that the filter of cofinite subsets of X is a subset of every nonprincipal ultrafilter on X. Show that it is in fact the intersection of all nonprincipal ultrafilters on X.*

EXERCISE 39

   (i) *If a filter is ultra, the corresponding quotient is the canonical two-valued boolean algebra $\{0, 1\}$.*
   (ii) *If $\mathcal{U}$ is a nonprincipal ultrafilter in $\mathcal{P}(I)$, then it contains all cofinite subsets of I. Deduce that if X is finite, all filters in $\mathcal{P}(X)$ are principal.*

EXERCISE 40 *An examination question*

   (i) *State Zorn's lemma.*
   (ii) *Let U be an arbitrary set and $\mathcal{P}(U)$ be the power set of U. For X a subset of $\mathcal{P}(U)$, the **dual** $X^{\vee}$ of X is the set $\{y \subseteq U : (\forall x \in X)(y \cap x \neq \emptyset)\}$.*
   (iii) *Is the function $\lambda X.X^{\vee}$ monotone? Comment.*
   (iv) *By considering the poset of those subsets of $\mathcal{P}(X)$ that are subsets of their duals, or otherwise, show that there are $X = X^{\vee}$.*
   (v) *What can you say about the fixed points of $\lambda X.X^{\vee}$ on the assumption that U is finite?*

4

Propositional calculus

So far I have been extremely careful not to say anything about languages that depends in any way on semantics. We are now going to introduce ourselves to two notions in logic that cannot, without perversity, be approached without semantics. They are **theory** – which is a kind of language, and a **logic** – which is a kind of theory.

If P is a propositional alphabet containing letters p_1, p_2 ..., then $\mathcal{L}(P)$ is to be the **language over** P: the set of all formulæ like $p_1 \vee p_2$, $p_3 \wedge \neg p_4$ and so on, all of whose literals come from P – as in section 2.2.1.

A **theory** is a set of formulæ closed under deduction, and members of this set are said to be **theorems** of the theory. What is deduction? This is where semantics enters. Rules of deduction are functions from tuples-of-formulæ to formulæ that preserve something, usually (and in the course of this book *exclusively*) truth.

But what is truth of a formula? A formula is a piece of syntax. It may be long or short, or ill-formed or well-formed. It can be true or false only with respect to an *interpretation*. Interpretations in the propositional calculus are simply rows from the things you may know and love as *truth-tables*: they are functions from literals to truth-values, to {true, false}. Each row in a truth-table is an interpretation of the formula. While we are about it, a **tautology** is a formula that is truth-table valid, that is, true under *all* interpretations.[1]

So a theory is a set of formulæ true in an interpretation or in a number of interpretations. If deductions are to be things that preserve truth, and

---

[1] This word is routinely misused. The other day my wife threatened to buy me a new pair of trousers so I said "I'd rather have the money instead" "That's a tautology" she said, thinking about the use of both 'rather' and 'instead'. She was wrong: it's not a tautology, it's a *pleonasm*.

truth is always truth-in-one-or-more-interpretations, then a theory will be a set of formulæ closed under deduction, as we wanted at the outset.

Here is an example of a propositional theory. We might call it the theory of adding two eight-bit words (with overflow). It has 24 propositional letters, p_0 to p_7, p_8 to p_{15} and p_{16} to p_{23}, and axioms to say that p_{16} to p_{23} represent the output of an addition if p_0 to p_7 and p_8 to p_{15} represent two words of input. true is 1 and false is 0, so it contains things like $((p_0 \wedge p_8) \rightarrow \neg p_{16})$ (because an odd plus an odd is an even!).

	p_7	p_6	p_5	p_4	p_3	p_2	p_1	p_0
+	p_{15}	p_{14}	p_{13}	p_{12}	p_{11}	p_{10}	p_9	p_8
=	p_{23}	p_{22}	p_{21}	p_{20}	p_{19}	p_{18}	p_{17}	p_{16}

A **logic** is a theory closed under **substitution**.

Before I tell you what substitution is, let's motivate it. The theory of adding two eight-bit words contains $((p_0 \wedge p_8) \rightarrow \neg p_{16})$ but not $((p_1 \wedge p_8) \rightarrow \neg p_{16})$, for example. It does not treat all propositional letters the same. This is because it is a description of a particular state of affairs rather than a general constraint on what states of affairs are possible. A set of formulæ that encapsulates a general constraint must make the same assertions about all propositional letters, must be impervious to differences between them and must be invariant under permutations that act on them. (Rather like the way in which – say – moral truths are invariant under permutations of moral agents: if it is wrong for me it is wrong for you. This is Kant's *categorical imperative*[1].) You should make a mental note here to understand that this is why logics are closed under substitution. It is an invariance property.[2]

Now let us be formal about it. A **substitution** is a (finite)[3] map from variables to formulæ (in the propositional case) or (in the predicate case) from variables to terms or from predicate letters to formulæ with the appropriate number of free variables. If L is a logic and σ is a substitution, then the result of applying σ to any formula in L must also be in L. I am going to assume that you know what I mean by applying

---

[1] I am indebted to Nick Denyer for showing me the Greek puzzle in Diogenes Laertius, *Lives of the Philosophers* Book 6, Chapter 97. "If Theodorus could not be said to be committing an injustice in doing something, then neither could Hipparchia be said to be committing an injustice in doing that thing. But Theodorus commits no injustice in hitting himself. So neither does Hipparchia commit an injustice in hitting Theodorus".

[2] See the posthumous article of Tarski (1986).

[3] We can probably drop this condition, since all the formulæ the substitution will act on are finite.

a substitution (which is a function defined on variables) to a formula. We will use the specific notation '$A[\phi/\psi]$' for the result of replacing in A all occurences of ψ by ϕ.

Now just as we were not interested in just any old language (= set of strings of letters) but only languages that look as if they are going to have some semantics, so we are not going to become interested in just any old set of propositional (or predicate) formulæ closed under substitution and something that looks like deduction but only in logics that are the set of formulæ whose burden is that it is legitimate to reason in a certain way (for example, $p \wedge q \rightarrow q$ tells us it is all right to infer q from $p \wedge q$) or the set of all formulae true in some (very large and natural) set of interpretations. For example, we will be very interested in that propositional logic that is the set of tautologies.

Formally, a **valuation** (or interpretation) is a function from propositional letters to truth-values. A valuation can be thought of as a conjunction of literals and negations of literals (i.e., as the conjunction of those literals that it believes to be true, and the negations of those literals that it believes to be false). Next we define `eval`: valuations × formulæ → truth-values by recursion on formulæ.

DEFINITION 15 *Let 'v' range over valuations.*[1]

$$
\begin{aligned}
\texttt{eval}(A, v) &:= & v(A), \text{ if } A \text{ is a literal.} \\
\texttt{eval}(A \wedge B, v) &:= & \texttt{eval}(A, v) \wedge \texttt{eval}(B, v). \\
\texttt{eval}(A \vee B, v) &:= & \texttt{eval}(A, v) \vee \texttt{eval}(B, v). \\
\texttt{eval}(A \rightarrow B, v) &:= & \texttt{eval}(A, v) \rightarrow \texttt{eval}(B, v). \\
\texttt{eval}(\neg A, v) &:= & \neg \texttt{eval}(A, v).
\end{aligned}
$$

This enables us to think of **satisfaction** as being a relation between formulæ and valuations. v **satisfies** A if $\texttt{eval}(A, v) = $ **true**. (In the propositional case, where we are at the moment, this sounds a bit obsessional, but thinking of it in this rather abstract way will help later when we come to semantics for predicate logic) The fact that `eval` is defined on all formulæ means that we can think

---

[1] To be strictly correct, one should add that the letter A is a variable ranging over formulæ. If you are unhappy about putting symbols like '$\wedge$' between names of formulæ instead of between formulæ, you may wish to look ahead to page 107. Alternatively, you might prefer to regard '$A \vee B$' as a special compound variable constrained to vary only over disjunctions, '$\neg A$' as a special compound variable constrained only to range over negations, and so on.

of a valuation as a complete description of a way the world (or at least the world as described by propositional logic) can be: via `eval`, a valuation determines the truth value of every formula.

To each formula ϕ we can associate a function from valuations to truth-values, namely the function that sends a valuation to `true` if the valuation satisfies ϕ and to `false` otherwise. That way we can think of any formula as the set of those valuations that make it true, and this pairing of formulæ with sets-of-valuations is 1-1 up to semantic equivalence. (Two formulæ are **semantically equivalent** if they are satisfied by the same valuations.) This enables us to think of a formula (or rather of formulæ-up-to-semantical-equivalence) as a function from valuations to truth-values, or as the disjunction of all the valuations that think it is true, so that any formula can be thought of as a disjunction of conjunctions of literals-and-negations-of-literals.

This is the **normal form theorem**:

THEOREM 16 *Every propositional formula is semantically equivalent to one in disjunctive normal form.*

There is a dual theorem that says that any formula can be thought of as a *conjunction* of *disjunctions* of literals-and-negations-of-literals, but it cannot be given the same slick proof and it is simplest to derive it from the disjunctive normal form theorem by the de Morgan laws below (definition 17).

EXERCISE 41 *Theorem 16 tells us that – up to logical equivalence – there are 2^{2^n} distinct propositional formulæ with n propositional letters. Put another way, this says that the free boolean algebra with n generators has 2^{2^n} elements. How many bases are there for the free boolean algebra with 2^{2^n} elements?*

This goes back to Boole, who "derived" it by using the Taylor-MacLaurin theorem over a boolean ring.

We can also prove by structural induction on formulæ that such a normal form can always be found.

Truth-tables enable us to discriminate between those formulæ that always come out true and those that might not. We call the first *valid*.

DEFINITION 17 *A valid formula is one that is true under all interpretations.*

I have carefully phrased this definition so it covers both propositional and predicate calculi. (I have not said anything yet about what an interpretation is in predicate calculus). Many tautologies have standard proper names: $A \vee \neg A$ is **excluded middle**, $\neg\neg A \rightarrow A$ is **double negation**, $\neg(A \wedge B) \longleftrightarrow (\neg A \vee \neg B)$ and $\neg(A \vee B) \longleftrightarrow (\neg A \wedge \neg B)$ are the **de Morgan laws** and $((A \rightarrow B) \rightarrow A) \rightarrow A$ is **Peirce's Law**.

The most important logic for us is classical logic: this logic contains all formulæ that are true under all interpretations.

EXERCISE 42 *Suppose A is a propositional formula and p is a letter appearing in A. Show that there are formulæ A_1 and A_2 not containing p such that A is semantically equivalent to $(A_1 \wedge p) \vee (A_2 \wedge \neg p)$.*

4.1 Semantic and syntactic entailment

I said earlier that deduction was a function on formulæ that preserved truth-in-an-interpretation, and we saw the two ways of thinking of theories. A theory will typically be the set of things true in some fixed interpretation or bundle of interpretations.

However, although that is the usual reason for interest in any specific theory, the theory itself might happen to be – and be conveniently studied *as* – a rectype. The idea that a body of truths (which is what a theory is supposed to be, after all) can be represented as an inductively generated set in this way goes back to Euclid. The rectype has axioms (like IN has 0) and rules of inference (like IN has the successor function).

Now we must think about the connection between theories as *semantically* characterised (set of things true in an interpretation) and theories *syntactically* characterised (things deducible from axioms). These two characterisations give rise to two relations between formulæ and sets of formulae.

We write "$\psi \models \phi$" to mean that any interpretation that satisfies ψ also satisfies ϕ. We overload '$\models$' by writing "$\Gamma \models \phi$", where Γ is a set of formulæ to mean that any interpretation satisfying all formulæ in Γ also satisfies ϕ. This is called **semantic entailment**.

We write "$\psi \vdash \phi$" to mean that ϕ follows from ψ by means of whatever the rules of inference are that we are using. These will typically be clear from the context. Again there is a version of this notation for sets of formulæ: "$\Gamma \vdash \phi$" means that ϕ can be deduced from assumptions in Γ. This is called **syntactic entailment**. We will write '$L, A \vdash \ldots$', where A is a single formula, to be the same as '$L \cup \{A\} \vdash \ldots$'.

The aim is to prove that these two notions are the same. "Of course", the astute reader will say, "this is trivial; just cook up the axioms and rules of inference so that they are". Not so: there are funny theories that cannot be expressed as rectypes in this way (see exercise 85 later).

DEFINITION 18 *A theory that is also a finitely presented rectype is said to be* **axiomatisable**.

If all one wanted to do was show that the set of propositional tautologies was a rectype (was an axiomatisable theory), the simplest thing to do would be to exhibit an axiomatisation and show that the set of things deducible from it is precisely the set of tautologies. However, I shall complicate matters by introducing not one but *two* rectypes of formulæ and showing that all three sets are the same.

We defined natural numbers as things one can obtain from 0 by adding 1 repeatedly. Any rectype is built up from founders by means of **operations** also known as **constructors**. With many rectypes there are alternative ways of generating its members.[1] Perhaps a lot of founders and very few operations, or a lot of operations and very few founders. This is certainly the case with the rectypes that constitute the logics we are interested in. We can either have many founders (axioms) and very few operations (rules of inference) (typically only *modus ponens*) or many rules of inference and few – if any – founders.

4.1.1 Many founders, few rules: the Hilbert approach

In the particular treatment here, we have only two primitive connectives, $\to$ and $\perp$; all others are defined in terms of them:

$\neg A$ is $A \to \perp$

$A \wedge B$ is $\neg(A \to \neg B)$

$A \vee B$ is $\neg A \to B$.

Here is one set of axioms.

K: $A \to (B \to A)$

S: $(A \to (B \to C)) \to ((A \to B) \to (A \to C))$

T: $(\neg B \to \neg A) \to ((\neg B \to A) \to B)$.

---

[1] Be careful not to confuse this with the situation where an element of a rectype can be generated in two ways **from the one set of rules** ("The thing that terrified him was climbing up the drainpipe"). I am alluding here to the situation where there are two different sets of rules for generating the same set. This situation may be familiar to you: a given group may have several different presentations.

The third axiom does not have a generally accepted proper name. The names of the first two axioms are motivated by the Curry-Howard correspondence.

But we also need rules that enable us to infer things from our axioms. These are a rule of substitution (every subsitution-instance of a theorem is a theorem) and the rule of **modus ponens**: from A and $A \to B$ infer B. Recall in this connection the habit first shown on page 26 of presenting an inference with premisses above and conclusion below the line. It is customary to display the *modus ponens* rule as:

$$\frac{A \quad A \to B}{B}.$$

Recall the idea of a proof from section 2.1.7. Naturally, the corresponding notion here is a lot more complicated, though it is quite easy to reconstruct what it must be. A **Gödel-style** proof of ψ from Γ is a finite list of formulæ ending with ψ wherein every formula is either a member of Γ or is obtained from an earlier member of the list by substitution or from two earlier items in the list by means of modus ponens.[1]

This definition of a Gödel-style proof makes proofs into things that are just as much mathematical objects as are numbers or groups or anything else. This is a distinctive development of twentieth century mathematics. Proofs are members of an inductively defined set: any list of substitution instances of axioms is a proof; any list obtained by appending on the end of a list l a formula obtained by doing *modus ponens* to two formulæ in l is a proof. A theorem is the last member of a proof.

EXERCISE 43 *Construct Gödel-proofs of the following:*

(a) $B \to \neg\neg B$.

(b) $\neg A \to (A \to B)$.

(c) $A \to (\neg B \to \neg(A \to B))$.

(d) $(A \to B) \to ((\neg A \to B) \to B)$.

(e) $\bot \to A$.

---

[1] "An argument isn't just contradiction, it is a reasoned series of steps tending to establish a conclusion." "No it *isn't!*" – the Blessed Python.

4.1.1.1 The deduction theorem

The *deduction theorem* for a logic L is the assertion

$$\text{if } L, A \vdash B, \text{ then } L \vdash A \to B.$$

The converse is easy.

THEOREM 19 *The deduction theorem holds for L iff L contains (all substitution instances of) K and S.*

Proof:

$L \to R$

The left-to-right direction is easy, for we can use the deduction theorem to construct proofs of K and S. This we do as follows:

$$L \vdash (A \to (B \to C)) \to ((A \to B) \to (A \to C))$$

(which is what we want) holds iff (by the deduction theorem)

$$L \cup \{(A \to (B \to C))\} \vdash ((A \to B) \to (A \to C))$$

iff (by the deduction theorem)

$$L \cup \{(A \to (B \to C)), (A \to B)\} \vdash (A \to C)$$

iff (by the deduction theorem)

$$L \cup \{(A \to (B \to C)), (A \to B), A\} \vdash C.$$

But this last one we can certainly do, since

$$[(A \to (B \to C)); (A \to B); A; (B \to C); B; C]$$

is a Gödel-proof of C from $L \cup \{(A \to (B \to C)), (A \to B), A\}$ (and we have already seen how to do this by natural deduction). We also want $L \vdash A \to (B \to A)$. This holds (by the deduction theorem) iff $L \cup \{A\} \vdash (B \to A)$ iff (by the deduction theorem again) $L \cup \{A, B\} \vdash A$.

$R \to L$

Suppose $L, A \vdash B$. That is to say, there is a (Gödel) proof of B in which A is allowed as an extra axiom. Let the ith member of this list be B_i. We prove by induction on i that $L \vdash A \to B_i$. $B_i \to (A \to B_i)$ is always a (substitution instance of) an axiom (because of K), so if B_i is an axiom, we have $L \vdash A \to B_i$ by *modus ponens*. If B_i is A, this follows because $L \vdash A \to A$. If B_i is obtained by *modus ponens* from two earlier things in the list, say B_j and $B_j \to B_i$, then by induction hypothesis we have $L \vdash A \to B_j$ and $L \vdash A \to (B_j \to B_i)$. But by

S this second formula gives us $L \vdash (A \to B_j) \to (A \to B_i)$ and then $L \vdash A \to B_i$ by *modus ponens*.   ∎

What the deduction theorem says is that a particular relation between formulæ (namely deducibility) is actually representable by a connective within the language to which the formulæ belong. As we have seen, K and S can be proved just using →-introduction and →-elimination, and what theorem 19 tells us is that any system with a connective that obeys *modus ponens* (and so behaves like a conditional) but does not obey →-introduction will not obey the deduction theorem. Logics set up to capture *intensional* concepts of implication typically fail to obey K. After all, one cannot expect to infer from A that there is a connection of meaning between B and A that is enough to deduce A from B: one would expect this to depend on B.

All readers should at least attempt exercise 43. It will bring home to them how difficult it is to construct proofs of tautologies from these axioms with substitution and *modus ponens* as sole rules of inference. If one is trying to prove B, then one has to find A such that both A and $A \to B$ can be proved. The problem is that there are infinitely many A's that are candidates for this rôle, with the result that there is on the face of it no feasible search strategy for proofs. Suppose we had a finite collection X of formulæ all arising somehow from B, such that whenever there is an A such that both $\vdash A \to B$ and $\vdash A$, then there is such an A in X, then we would have a procedure for reliably finding proofs.

The solution is to have few founders but a lot of rules, but let us not leap into it without a bit of motivation. Anyone who has tried proving theorems from these axioms will not only have noticed how difficult it is, but will also have spotted how useful the deduction theorem is. One is tempted to describe the deduction theorem as a *derived rule of inference*, but of course it is nothing of the kind. It does not provide proofs in the system, but it does provide (meta)proofs that such proofs can be found. (And it does so *constructively*: a (meta)proof that there is a proof of B can be teased apart to furnish a proof of B.) However, if we are to procede from many-founders-and-few-rules to many-rules-but-few-founders, the usefulness of the deduction theorem gives us strong hints about what those rules should be.

EXERCISE 44 *Let T be an axiomatisable theory and ψ an arbitrary theorem of T (that is not a truth-table tautology). Show that T has an* axiomatisation $A \cup \{\psi\}$ where ψ does not follow from A. *(Hint: Use Peirce's law.)*

4.1.2 No founders, many rules

We are now going to construct the same rectype of formulæ as before, this time with many rules (constructors) and no founders (axioms). To be more precise, what we are going to do in this treatment is define a rather more complex rectype of *proofs* and then recover a theory from it by discarding all the proofs but retaining what they were proofs of. This style of proof system is known as **natural deduction**.

In the following table we see that for each connective we have two rules: one to introduce the connective and one to eliminate it. (Some of these rules are bifid.)

$$\vee\text{-int: } \frac{A}{A \vee B}; \quad \frac{B}{A \vee B}; \qquad \vee\text{-elim: } \frac{A \vee B \quad \overset{[A]}{\underset{\vdots}{C}} \quad \overset{[B]}{\underset{\vdots}{C}}}{C}$$

$$\wedge\text{-int: } \frac{A \quad B}{A \wedge B}; \qquad\qquad \wedge\text{-elim: } \frac{A \wedge B}{A}; \quad \frac{A \wedge B}{B}$$

$$\rightarrow\text{-int } \frac{\overset{[A]}{\underset{\vdots}{B}}}{A \rightarrow B}; \qquad\qquad \rightarrow\text{-elim: } \frac{A \quad A \rightarrow B}{B}$$

$$\text{Ex falso sequitur quodlibet; } \frac{\bot}{A} \quad \text{contradiction } \frac{\overset{[\neg A]}{\underset{\vdots}{\bot}}}{A}$$

These last two rules are the only rules that specifically mention negation. $\neg B$ is $B \rightarrow \bot$.

It is customary to connect the several occurrences of a single formula at introductions (it may be introduced several times) with its occurrences at elimination by means of superscripts. Square brackets are placed round eliminated formulæ.

Proofs are built up by assembling instances of these rules. Formulæ that appear as the bottom lines of proofs without live premises are theorems.

EXERCISE 45 *Find natural deduction proofs of the following formulæ:*

$(A \to (B \to C)) \to (B \to (A \to C))$.

$A \to (B \land C) \to (A \to B) \land (A \to C)$.

$((A \land B) \to C) \to A \to (B \to C)$.

$A \to (B \to A)$.

$A \to ((A \to B) \to B)$.

$(A \to (B \to C)) \to ((A \to B) \to (A \to C))$.

$(A \to B) \to (((A \to B) \to B) \to B)$.

$(((A \to B) \to B) \to B) \to A \to B$.

$((A \to B) \to A) \to ((A \to B) \to B)$.

$(A \land (B \lor C)) \to ((A \land B) \lor (A \land C))$.

EXERCISE 46 *Go back to section 4.1.1 for the moment, where we had only two primitives, $\to$ and $\neg$. Justify the introduction and elimination rules for the other connectives as derived rules, and verify that these connectives are symmetrical.*

These rules involve *action at a distance* in the following sense. Let us attempt to prove $(A \to (B \to C)) \to ((A \to B) \to (A \to C))$. The obvious thing to try – indeed the *only* thing to try – is the $\to$-introduction rule. We must have deduced $(A \to B) \to (A \to C)$ from $A \to (B \to C)$. So we know so far that our proof looks like

$$[A \to (B \to C)]$$

$$\vdots$$

$$\frac{(A \to B) \to (A \to C)}{(A \to (B \to C)) \to ((A \to B) \to (A \to C))}$$

The presence of the dots means not only that we do not at this stage know what the rest of the proof will be, it also means that we do not know how much space to leave for the bits that are to come! It is true that backward proof search is easier with natural deduction systems than with the Hilbert-style systems of the previous section, in that we have solved the problem of the unbounded search, but evidently not all the problems have disappeared. If we write '$\Gamma \vdash \psi$' to mean that there is a deduction of ψ from Γ, and hang on to the habit of writing premisses above and conclusions below the line, we can represent the information in the above illustration more economically by some picture like the following:

$$\frac{A \to (B \to C) \vdash (A \to B) \to (A \to C))}{\vdash (A \to (B \to C)) \to ((A \to B) \to (A \to C))}.$$

This says that if the upper assertion about the existence of a proof is correct, then so is the lower one.

This '⊢' symbol is of course not a symbol of the language of propositional logic. This is our first encounter with the **object language – metalanguage** distinction. The subject matter of a metalanguage is a language: thus $A \vdash B$ is an assertion (in a metalanguage) about the (object) language in which A and B appear, namely, propositional logic; $A \to B$ is an assertion *in* that object language. The advantage of reasoning about ⊢ is that it gives us an elegant programming solution to the problem of reasoning about proofs in propositional logic. Although the invention of sequent calculus antedates the invention of computing machinery by a decade, and antedates the development of theorem-proving by machine by several decades, there is merit in the anachronistic view that sequent calculus is the programming solution to the problem of backward search for proofs.

4.1.3 Sequent calculus

A **sequent** is a formula $\Gamma \vdash \psi$ where Γ is a set of formulæ and ψ is a formula. We know what this means: it means that there is a deduction of ψ from Γ. In sequent calculus one reasons about sequents rather than about the formulæ that compose them, as one did with natural deduction.

This definition is the one that arises from the idea of sequents-as-the-solution-to-programming-backward-search. However, it turns out that a more useful definition has to the right of the '⊢' not a single formula but a whole set of formulæ, as on the left. Try thinking of a sequent as saying that there is a proof of something on the right using only premisses found on the left.

The rules that this picture gives rise to are as follows:

$$\lor L : \frac{\Gamma, \psi \vdash \Delta \quad \Gamma, \phi \vdash \Delta}{\Gamma, \psi \lor \phi \vdash \Delta} \qquad \lor R : \frac{\Gamma \vdash \Delta, \psi, \phi}{\Gamma \vdash \Delta, \psi \lor \phi}$$

$$\land L : \frac{\Gamma, \psi, \phi \vdash \Delta}{\Gamma, \psi \land \phi \vdash \Delta} \qquad \land R : \frac{\Gamma \vdash \Delta, \psi \quad \Gamma \vdash \Delta, \phi}{\Gamma \vdash \Delta, \psi \land \phi}$$

$$\neg L : \frac{\Gamma \vdash \Delta, \psi}{\Gamma, \neg \psi \vdash \Delta} \qquad \neg R : \frac{\Gamma, \psi \vdash \Delta}{\Gamma \vdash \Delta, \neg \psi}$$

$\rightarrow L: \quad \dfrac{\Gamma \vdash \Delta, \phi \quad \Gamma, \psi \vdash \Delta}{\Gamma, \phi \rightarrow \psi \vdash \Delta} \qquad \rightarrow R: \quad \dfrac{\Gamma, \psi \vdash \Delta, \phi}{\Gamma \vdash \Delta, \psi \rightarrow \phi}$

and weakening: $\dfrac{\Gamma \vdash \Delta}{\Gamma, A \vdash \Delta, B}$;

contraction-L $\dfrac{\Gamma, A, A \vdash \Delta}{\Gamma, A \vdash \Delta}$; contraction-R $\dfrac{\Gamma \vdash \Delta, A, A}{\Gamma \vdash \Delta, A}$;

and cut: $\dfrac{\Gamma \vdash \Delta, A \quad \Gamma', A \vdash \Delta'}{\Gamma \cup \Gamma' \vdash \Delta, \Delta'}$.

There is no rule for the biconditional: we think of it as a conjunction of two conditionals.

We accept any sequent that has a formula appearing on both sides. Such sequents are called **initial sequents**.

Recall that we started off thinking of a sequent as saying that there is a proof of something on the right using only premisses found on the left. To illustrate, think about the rule ∧-L. It tells us we can infer "$A \wedge B \vdash C$" from "$A, B \vdash C$". Now "$A, B \vdash C$" says that there is a deduction of C from A and B. But if there is a deduction of C from A and B, then there is certainly a deduction of C from $A \wedge B$, because one can get A and B from $A \wedge B$ by two uses of ∧-elim. The →-L rule can benefit from some explanation as well. Assume the two sequents above the line. We want to use them to show that there is a derivation of something in Δ from $\phi \rightarrow \psi$ and all the premisses in Γ. The first sequent above the line tells us that there is either a deduction of something in Δ using premisses in Γ (in which case we are done) or there is a deduction of ϕ. But we have $\phi \rightarrow \psi$, so we now have ψ. But then the second sequent above the line tells us that we can infer something in Δ.

$A \vdash A$ is an initial sequent. Use ¬-R to infer $\vdash A, \neg A$. Now it just isn't true that there is always a proof of A or a proof of $\neg A$, so this example shows that it similarly just isn't true that a sequent can be taken to assert that there is a proof of something on the right using only premisses found on the left – unless we restrict matters so that there is only one formula on the right. However, it does help inculcate the good habit of thinking of sequents as metaformulæ, as things that formalise facts about formulæ rather than facts of the kind formalised by the formulæ.

no one is suggesting that sequent calculus is the right way to do the theory of proofs: there are too many obvious infelicities in its development. One rather glaring one is the fact that there is one obvious natural-deduction proof of $(A \vee (B \wedge C)) \rightarrow ((A \vee B) \wedge (A \wedge C))$, but there are two sequent versions of this rather than one. Miniexercise: Find these three proofs. Sequent calculus is now more than 70 years old, and modern proof theorists have developed a plethora of more subtle and complicated constructs in their attempts to better capture the underlying mathematics. This is an active area of research.

It is natural to think that the rules we use might have been chosen because there is a salient feature that they all preserve in the sense that, for each rule, if its inputs have that feature, so do its outputs. This is true: the rules preserve truth. (They also preserve validity.) But there are other properties that the rules might preserve, and consideration of them leads to weaker logics that are of some concern to logicians interested in computation.

Let us now drop down a level and return from sequents (assertions about inferences in the logic) to the logic itself. The logic we have just seen is called **classical logic**, and its inferences preserve truth. Preserving truth is **extensional** in that what is preserved is a property of what the thing proved evaluates to, evaluations being functions that take intensions to extensions. Some other logics preserve intensional properties. One interesting case is constructive logic, where what is preserved is not a property of the thing proved but rather a property of the set of available proofs of the thing proved. Rules of inference are thought of as operations on proofs, giving rise to new proofs. Constructive logic allows only those operations that preserve the property (of proofs) of *corresponding to a construction*. I have not said what a construction is exactly. One reason for this is that nobody has yet given a mathematically satisfactory account of it, but another reason is that the idea can be conveyed even without formalisation. The following (admittedly rather artificial) challenge makes the point very well.

Find x and y, both irrationals but both real, such that x^y is rational.

Well, we all know that $\sqrt{2}$ is irrational, so if $\sqrt{2}^{\sqrt{2}}$ is rational, we can take both x and y to be $\sqrt{2}$. On the other hand, if $\sqrt{2}^{\sqrt{2}}$ is *not* rational, we take x to be $\sqrt{2}^{\sqrt{2}}$ and y to be $\sqrt{2}$, and we then find that $x^y = 2$. So either way we succeed.

Except that we don't. The challenge was to *find* such a pair x and y, not merely to prove that such a pair exists. Contrast this with the proof of Cantor's theorem (theorem 6), where one has an algorithm that accepts a candidate injection and explicitly provides something not in its range.

There is something particularly appealing about an existence proof that can be easily transformed into a construction of the thing whose existence has been proved. Constructivists say that such proofs have the **existence property**.

A moment's reflection will make it clear why our proof of the existence of a pair does not have the existence property. The existence property will fail if at any stage in the proof we are in one of two cases, but do not know which, *but we nevertheless exploit the knowledge that we are in one of the two cases.* We must never exploit our knowledge that $A \lor B$ unless we also know A, or we know B. To preserve the existence property, we must ensure that whenever $\vdash A \lor B$ then $\vdash A$ or $\vdash B$. (We will see proofs with the same features in exercises 6.6.88 and 6.6.90.) This means that we may make no use of the law of the excluded middle.

A logic designed to respect these constraints is therefore developed not to capture the set of those inferences that preserve a nice property of formulæ, but to ensure that proofs in it have a nice property. This fits well with this section's presentation of theories as arising from rectypes of proofs. It is possible to give a Hilbert-style presentation for constructive logic, but it is much less natural.

I am not going to tell you what set of formulæ constructive logic regards as valid. As it happens, most of the pruning that needs to be done can be achieved by the simple device of requiring all our sequents to have only one formula on the right. It is not entirely clear why this is the case, but this restriction does at least ensure that the view of sequents as metaformulæ that say "there is a proof of something on the right using only premisses that appear on the left" is correct.

We can generalise the concept of valuation to include all functions from literals to – well, anything with the same signature as boolean algebras. (There must be operations to interpret '$\land$', '$\lor$' etc). The obvious candidate for such a structure would be a boolean algebra. However, enlarging the set of valuations in this way has no effect on the class of sentences certified as valid as long as the set of values of the valuations forms a boolean algebra. This gives us a way of characterising boolean algebras.

EXERCISE 47 *Show that a structure for the language of boolean algebras (i.e., with 0, 1, $\land$, $\lor$ and $\neg$) is a boolean algebra iff it validates all truth-table tautologies.*

So boolean algebras characterise classical logic. Is there a different kind of algebra that characterises constructive logic? Yes, there is, and these algebras are called **Heyting algebras**. A Heyting algebra is a complete distributive lattice, typically presented with a defined operator $\to$ where $p \to q$ is $\bigvee\{r \,:\, p \land r \leq q\}$. This is another example of overloading, for the arrow has already been used for the material conditional. Naturally this is deliberate. Since everything that is constructively correct is classically valid, but not vice versa, there must be algebras that are Heyting algebras but are not boolean algebras. In fact, there are plenty and, fortunately for people attempting this next exercise, some of them are very small.

EXERCISE 48

(i) *Show that K and S are satisfied in any Heyting algebra.*

(ii) *Show that $((A \to B) \to A) \to A$ (Peirce's law) cannot be deduced from K and S.*

Clearly one of the things that gives us trouble with $\sqrt{2}^{\sqrt{2}}$ is excluded middle (we *must* have the disjunction property if we are to have the existence property!). However, it does not mean that constructive logic thinks there are more than two truth-values. Quite the reverse!

EXERCISE 49 *Find a sequent calculus proof of*

$$\neg(A \longleftrightarrow B), \neg(A \longleftrightarrow C), \neg(C \longleftrightarrow B) \vdash$$

satisfying the single-conclusion constraint. (This is hard, and the proof is long!)

4.2 The completeness theorem

The axiomatic and the natural deduction approach both give rise to a notion of syntactic entailment. I shall show that both of these are the same as semantic entailment. This is the **completeness theorem**. First a toy completeness theorem. (Lesniewski 1929).

This is going to be sketched, to give you a taste of how these things work.

Pure biconditional logic has one connective, "$\longleftrightarrow$", and one propositional constant symbol $\bot$. There are three axioms:

$p \longleftrightarrow p;$

$(p \longleftrightarrow q) \longleftrightarrow (q \longleftrightarrow p);$
$((p \longleftrightarrow q) \longleftrightarrow r) \longleftrightarrow (p \longleftrightarrow (q \longleftrightarrow r));$

We do not have a negation sign, but if we want $\neg p$, we can introduce it as $p \longleftrightarrow \bot$.

We also have a rule of *modus ponens* and a rule of substitutivity of the biconditional: if A and $\phi \longleftrightarrow \psi$, then $A[\phi/\psi]$. (Recall that $A[\phi/\psi]$ is the result of replacing in A all occurences of ψ by ϕ.)

We are going to show that something is a (truth-table) valid expression of this logic iff it is derivable from these axioms. In fact we can show

EXERCISE 50 *The following are equivalent:*
ϕ *is valid.*
ϕ *is a consequence of the three above axioms.*
Every propositional letter appearing in ϕ appears an even number of times.

First we prove that for any two formulæ ϕ and ψ with the same multiset of literals, we have $\psi \vdash \phi$ and $\phi \vdash \psi$. (We say ϕ and ψ are **interdeducible**.) Then, if Φ has two occurrences of p, it will be interdeducible with something of the form $(p \longleftrightarrow p) \longleftrightarrow \Phi'$.

■

Now we return to the main plot: the completeness theorem for propositional logic.

THEOREM 20 *The completeness theorem for propositional logic. The following are equivalent:*
(1) ϕ is provable by natural deduction.
(2) ϕ is provable from the three axioms K, S and T.
(3) ϕ is truth-table valid.

Proof: We will prove that $(3) \to (2) \to (1) \to (3)$.

$(2) \to (1)$ First we show that all our axioms follow by natural deduction – by inspection. Then we use induction: if there are natural deduction proofs of A and $A \to B$, there is a natural deduction proof of B!

$(1) \to (3)$ To show that everything proved by natural deduction is truth-table valid we need only note that, for each rule, if the hypotheses are true (under a given valuation), then the conclusion is too. By induction on composition of rules this is true for molecular proofs as well.

If we have a molecular proof with *no* hypotheses, then vacuously they are all true (under a given valuation), so the conclusion likewise is true (under a given valuation). But the given valuation was arbitrary, so the conclusion is true under all valuations.

(3) → (2) (This proof is due to Mendelson (1979).) Now to show that all tautologies follow from Mendelson's axioms.

At this point we must invoke exercise 43, since we need the answers to complete the proof of this theorem. It enjoins us to prove the following:

(a) $B \to \neg\neg B$.
(b) $\neg A \to (A \to B)$.
(c) $A \to (\neg B \to \neg(A \to B))$.
(d) $(A \to B) \to ((\neg A \to B) \to B)$.

If we think of a propositional formula in connection with a truth-table for it, it is natural to say things like: $p \longleftrightarrow q$ is true as long as p and q are both true or both false, and false otherwise. Thus truth-tables for formulæ should suggest to us deduction relations like

$$A, B \vdash A \longleftrightarrow B,$$

$$\neg A, \neg B \vdash A \longleftrightarrow B,$$

and similarly

$$A, \neg B \vdash \neg(A \longleftrightarrow B).$$

To be precise, we can show:

Let A be a molecular wff containing propositional letters $p_1 \ldots p_n$, and let f be a map from $\{k \in \mathbb{N} : 1 \leq k \leq n\}$ to $\{\texttt{true}, \texttt{false}\}$. If A is satisfied in the row of the truth-table where p_i is assigned truth-value $f(i)$, then

$$P_1 \ldots P_n \vdash A,$$

where P_i is p_i if $f(i) = \texttt{true}$ and $\neg p_i$ if $f(i) = \texttt{false}$. If A is not satisfied in that row, then

$$P_1 \ldots P_n \vdash \neg A,$$

and we prove this by a straightforward induction on the rectype of formulæ.

We have only two primitive connectives, $\neg$ and $\to$, so two cases. $\neg$

Let A be $\neg B$. If B takes the value $\texttt{true}$ in the row $P_1 \ldots P_n$, then, by the induction hypothesis, $P_1 \ldots P_n \vdash B$. Then, since $\vdash p \to \neg\neg p$ (this is exercise 43(a)), we have $P_1 \ldots P_n \vdash \neg\neg B$, which is to say $P_1 \ldots P_n \vdash \neg A$, as desired. If B takes the value $\texttt{false}$ in the row $P_1 \ldots P_n$, then, by the induction hypothesis, $P_1 \ldots P_n \vdash \neg B$. But $\neg B$ is A, so $P_1 \ldots P_n \vdash A$.

$\to$

Let A be $B \to C$.

Case (1): B takes the value $\texttt{false}$ in row $P_1 \ldots P_n$. If B takes the value $\texttt{false}$ in row $P_1 \ldots P_n$, then A takes value $\texttt{true}$ and we want $P_1 \ldots P_n \vdash A$. By the induction hypothesis we have $P_1 \ldots P_n \vdash \neg B$. Since $\vdash \neg p \to (p \to q)$ (this is exercise 43(b)), we have $P_1 \ldots P_n \vdash B \to C$, which is $P_1 \ldots P_n \vdash A$.

Case (2): C takes the value $\texttt{true}$ in row $P_1 \ldots P_n$. Since C takes the value T in row $P_1 \ldots P_n$, A takes value $\texttt{true}$, and we want $P_1 \ldots P_n \vdash A$. By the induction hypothesis we have $P_1 \ldots P_n \vdash C$, and so, by K, $P_1 \ldots P_n \vdash B \to C$, which is to say $P_1 \ldots P_n \vdash A$.

Case (3): B takes value $\texttt{true}$ and C takes value $\texttt{false}$ in row $P_1 \ldots P_n$. A therefore takes value $\texttt{false}$ in this row, and we want $P_1 \ldots P_n \vdash \neg A$. By the induction hypothesis we have $P_1 \ldots P_n \vdash B$ and $P_1 \ldots P_n \vdash \neg C$. But $p \to (\neg q \to \neg(p \to q))$ is a theorem (this is exercise 43(c)) so we have $P_1 \ldots P_n \vdash \neg(B \to C)$, which is $P_1 \ldots P_n \vdash \neg A$.

Suppose now that A is a formula that is truth-table valid and that it has propositional letters $p_1 \ldots p_n$. Then, for example, both $P_1 \ldots P_{n-1}$, $p_n \vdash A$ and $P_1 \ldots P_{n-1}, \neg p_n \vdash A$, where the capital letters indicate an arbitrary choice of $\neg$ or null prefix as before. So, by the deduction theorem, both p_n and $\neg p_n \vdash (P_1 \wedge P_2 \ldots \wedge P_{n-1}) \to A$ and we can certainly show that $(p \to q) \to (\neg p \to q) \to q$ is a theorem (this is exercise 43(d)), so we have $P_1 \ldots P_{n-1} \vdash A$, and we have peeled off one hypothesis. Clearly this process can be repeated as often as desired to obtain $\vdash A$. ∎

The following equivalent assertion – and its analogue for predicate logic – is known as the completeness theorem as well and is sometimes a more useful formulation.

COROLLARY 21 ϕ *is consistent (not refutable from the axioms) iff there is a valuation satisfying it.*

Proof: $\nvdash \neg\phi$ (i.e., ϕ is consistent) iff $\neg\phi$ is not tautologous. This is turn is the same as ϕ being satisfiable. ∎

If $T_1 \subseteq T_2$ are theories, we say that T_2 is an **extension** of T_1. If $T_1 \neq T_2$, then T_2 is a **proper** extension. (I warned you the word 'extension' would be overloaded!) A theory with no consistent proper extension is – reasonably enough – said to be **complete**. Beware: The completeness theorem is not so-called because it says that the set of all tautologies is a complete theory: it isn't!

4.2.1 Lindenbaum algebras

The Lindenbaum algebra of a theory T is the set of T-interdeducibility classes of formulæ partially ordered by deducibility. (We now know that semantic and syntactic entailment are the same, so it does not matter which we mean.) That is to say, if $[\phi]$ and $[\psi]$ are the equivalence classes of ϕ and ψ, respectively, then $[\phi] \leq [\psi]$ if ϕ (or anything T-interdeducible with it) $\rightarrow \psi$ (or anything interdeducible with it). The complement of $[\psi]$ is naturally $[\neg\psi]$. The Lindenbaum algebra is a boolean algebra as long as T contains all (propositional) tautologies.

The Lindenbaum algebra of the empty theory over an alphabet is of course the free boolean algebra generated by the literals of that alphabet. Any theory T over that alphabet corresponds to a filter in this algebra, and T is consistent iff this filter is proper.

The filter generated by a set of points in the Lindenbaum algebra of a theory T is just the theory axiomatised by T plus those axioms:

(i) If T' is a theory extending T, then the set of equivalence classes of theorems of T' form a filter in the Lindenbaum algebra of T.

(ii) If T' is a theory extending T, then the Lindenbaum algebra of T' is isomorphic to the quotient algebra modulo the filter of the previous remark.

(iii) If T' is a *complete* extension of T, then the corresponding filter is ultra.

4.2.2 The compactness theorem

The compactness theorem strictly is an assertion to the effect that a certain topology on the space of all interpretations is compact. That is how it got its name. In fact, most logic texts make nothing of this fact. One that does is Peter Johnstone's book [1987]. Look at exercise 2.6 on page 17.

There are two rather different-sounding facts that are both known as the compactness theorem.

THEOREM 22

(i) *If T is a theory such that every finite subset of T has an interpretation making it true, then T has such an interpretation.*

(ii) *Every consistent theory has a complete consistent extension.*

Proof:

(i) If every finite subset of T has an interpretation, then every finite subset of T is consistent and does not imply a contradiction. So there can be no proof of a contradiction in T either, because any proof of a contradiction would be finite and would appear in one of the finite subsets of T – which all have interpretations and so are consistent. But corollary 21 tells us that if T is consistent, it has an interpretation.

We can prove (ii) by means of Zorn's lemma. However, it can also be proved (more economically) by reasoning about Lindenbaum algebras and using the prime ideal theorem (theorem 12). On page 89 we saw how filters in the Lindenbaum algebra of a theory T correspond to extensions of T. The prime ideal theorem tells us that there is an ultrafilter in the Lindenbaum algebra of T. This ultrafilter corresponds to a complete extension of T. ∎

There are various applications of the prime ideal theorem, and the one that follows is fairly typical in that it uses the prime ideal theorem to "glue" together partial solutions to a problem.

A k-colouring of a graph is a map assigning to each vertex one of k colours in such a way that adjacent vertices are given different colours.

REMARK 23 *If every finite subgraph of a graph is n-colourable, then the graph itself is n-colourable.*

Proof: I am going to take the case $n = 4$, for ease of illustration. We have an infinite graph $\langle G, E \rangle$, all of whose finite subgraphs are 4-colourable.

The language: For each vertex x we have four propositional letters p_x, q_x, r_x, s_x. Each of these corresponds to an assertion that x has been painted some given colour. For each pair x, y of vertices we have a propositional letter $c_{x,y}$, which will be used to say whether or not x and y are connected.

(Notice that the 'x' is not a variable and that, strictly speaking, these propositional letters – 'p_x', 'q_y', '$c_{x,y}$', and so on – have *no internal structure* and the only reason why we write them out like this is to make it more obvious what is going on. The subscripts are not even part of the *syntax* but merely part of the *typesetting*! The information coded by the subscripts is not preserved by relettering of variables, which is a fairly mild process that ought to preserve anything of interest.)

The theory: We adopt the following axiom schemes:

(i) $(p_x \wedge \neg q_x \wedge \neg r_x \wedge \neg s_x) \vee (\neg p_x \wedge q_x \wedge \neg r_x \wedge \neg s_x) \vee (\neg p_x \wedge \neg q_x \wedge r_x \wedge \neg s_x) \vee (\neg p_x \wedge \neg q_x \wedge \neg r_x \wedge s_x)$ for each x;

(ii) $c_{x,y} \rightarrow \neg(p_x \wedge p_y) \wedge \neg(q_x \wedge q_y) \wedge \neg(r_x \wedge r_y) \wedge \neg(s_x \wedge s_y)$ for each x and y;

(iii) $c_{x,y}$ if G has an edge joining x and y, and $\neg c_{x,y}$ if not.

The first scheme says that every vertex has precisely one colour. The second says that adjacent vertices have different colours.

This theory is consistent since all its finite subtheories are consistent. The only thing they can say is that some finite subgraph is 4-colourable, and we are told that this is true.

So there is a valuation v making all these axioms true. Any such valuation gives rise to a 4-colouring of G: if $v(p_x) = \texttt{true}$, we colour vertex x with colour p, and so on.

EXERCISE 51 *Prove the order extension principle (page 62) using the compactness theorem for propositional logic instead of Zorn's lemma.*

4.2.2.1 Interpolation

Recall that $\mathcal{L}(P)$ is the set of propositional formulæ that can be built up from the literals in the alphabet P. Let us overload this notation by letting $\mathcal{L}(s)$ be the set of propositional formulæ that can be built up from the literals in the formula s.

Suppose $s \rightarrow t$ is a tautology, but $\mathcal{L}(s) \cap \mathcal{L}(t) = \emptyset$. What can we say? Well, there is no valuation making s true and t false. But, since valuations of s and t can be done independently, it means that either there is no valuation making s true, or there is no valuation making t false. With a view to prompt generalisation, we can tell ourselves that even if s and t have no letters in common, $\mathcal{L}(s) \cap \mathcal{L}(t) \neq \emptyset$, because $\texttt{true}$ is the conjunction of the empty set of formulæ and $\texttt{false}$ is the disjunction of the empty set of formulæ and that what we have proved is that either $s \rightarrow \texttt{false}$ is a tautology (so s is the negation of a tautology) or $\texttt{true} \rightarrow t$ is a tautology (so t is a tautology). But since $s \rightarrow \texttt{true}$ and $\texttt{false} \rightarrow t$ are always tautologies, we can tell ourselves that what we have established is that there is some formula ϕ in the common vocabulary (which must be either $\texttt{true}$ or $\texttt{false}$) such that both $s \rightarrow \phi$

and $\phi \to t$ are tautologies. If we now think about how to do this "with parameters" we get a rather more substantial result.

THEOREM 24 *(The interpolation lemma) Let P, Q, and R be three disjoint propositional alphabets; let s be a formula in $\mathcal{L}(P \cup Q)$ and t a formula in $\mathcal{L}(Q \cup R)$. If $s \vdash t$, then there is $u \in \mathcal{L}(Q)$ such that $s \vdash u$ and $u \vdash t$.*

Proof: We do this by induction on the number of variables common to s and t. We have already established the base case, where $\mathcal{L}(s) \cap \mathcal{L}(t)$ is empty. Suppose now that s and t have $n + 1$ variables in common. Let the $n + 1$th be 'p'. Then there are s' and s'', both p-free, such that s is equivalent to $(s' \wedge p) \vee (s'' \wedge \neg p)$ (this is by exercise 42). Similarly, there are t' and t'' such that t is equivalent to $(t' \wedge p) \vee (t'' \wedge \neg p)$. We know that any valuation making s true must make t true. But, also, any valuation making s true must either make $s' \wedge p$ true (in which case it makes $t' \wedge p$ true) or make $s'' \wedge \neg p$ true (in which case it makes $t'' \wedge \neg p$ true). So $s' \vdash t'$ and $s'' \vdash t''$. By induction there are interpolants u' and u'' such that $s \vdash u'$, $u' \vdash t'$, $s'' \vdash u''$ and $u'' \vdash t''$. The interpolant we need for s and t is $(u' \wedge p) \vee (u'' \wedge \neg p)$.   ∎

4.2.3 Nonmonotonic reasoning

In artificial intelligence there are people who are interested in what they call **nonmonotonic reasoning**, which is an attempt to formalise inferences like the following:

Tweety is a bird; we haven't yet been told that Tweety cannot fly. Accordingly deduce: Tweety can fly.

The reason why this kind of reasoning is called 'nonmonotonic' is that the set of propositions that get established by its methods is not a monotonically $\subseteq$-increasing function of time. After all, we might discover later that Tweety is a Penguin and then some hasty retractions will be in order, since we presumably have a rule that penguins cannot fly. In contrast, building up a stock of conclusions by means of the rules of inference we have seen is, indeed, monotonic. Consider the operation that takes a set Γ of formulæ and returns $\Gamma \cup$ the set of all formulæ Q such that $P \to Q$ and Q are in Γ. This is clearly a monotone function on the power set of the set of all formulæ, and there is no problem in showing that it will have a fixed point which will be the deductive closure of Γ. If you have rules of inference that say "if you believe this but don't believe that, then resolve to believe the other"[1] then you cannot rely on theorem 8 to tell you there are fixed points/deductive closures. In this connection look at question 3.1.3.24(ii). I hope I don't have to emphasise that nonmonotonic reasoning is a mess!

4.3 Exercises on propositional logic

(i)   (a) A *monotone* propositional function is one that will output 1 if all its inputs are 1. Show that no nonmonotone function can be defined in terms of any set of monotone functions. (easy)

(b) Show that NAND and NOR cannot be constructed by using ∧ and ∨ and → alone.

(c) Show that none of NAND, NOR, →, ∧, ∨ can be constructed by using XOR alone.

(d) Show that XOR and ⟷ and ¬ cannot be defined from ∨ and ∧ alone.

(e) Which propositional connectives can be defined in terms of the ternary connective if p then q else r?

(ii) What is a boolean algebra? Find a natural partial order on the set of binary connectives that makes them into a boolean algebra.

(iii) How many truth-functions of three propositional letters are there? Of four? Of n?

(iv) Prove that $\mathcal{P}(\{0, 1, 2\})$ and $\{T, \bot\}^3$ are isomorphic posets.

---

[1] Remember that withholding belief from p is not the same as according belief to ¬p!!

5

Predicate calculus

5.1 The birth of model theory

An important spur to the development of logic was the problem of the axiom of parallels and the discovery of noneuclidean geometry. If we are trying to determine whether or not the axiom of parallels follows from Euclid's other axioms, what do we do? If it does, life is easy, for it will be sufficient to exhibit a proof. If it does not, we need to demonstrate that there is no proof.

One way to do that would be to develop a formal device of proof-as-mathematical-object and then show that every proof fails to be a proof of the parallel axiom from the other axioms. Although we have made a start on proofs-as-mathematical-objects, we have not yet come up with anything that has been much good for constructing independence results.

Another possible approach would be to exhibit a model universe in which the parallel axiom is false but all the other axioms of Euclid are true. This requires us analogously to have a formal device of model-as-mathematical-object. For this we have to develop a robust concept of a *formula being true in a structure*. That is to say, we need **semantics**. We have already heard the phrase: *the autonomy of syntax*. Independence proofs of this second kind become possible only once one has started thinking of a syntax as something entirely separate from the subject matter it was devised to denote. The most characteristic products of twentieth-century logic, the completeness theorems, arise like the Greek conception of sexuality from the need to rejoin the two halves of this beast. This approach has been much more fruitful than the approach through proofs: we have a much better concept of model than of proof.

A completeness theorem is something that identifies a syntactic property of a formula (like having an even number of occurrences of every variable) with a semantic property (like being true under all interpretations). The major result of this chapter will be the completeness theorem for predicate calculus: we will exhibit a recursive datatype of formulæ (the syntactic characterisation) that is precisely the set of those formulæ that are true in all interpretations (the semantic characterisation).

5.2 The language of predicate logic

EXERCISE 52 *Fix a language $\mathcal{L}$ with constants and function letters (see section 2.2.2). A substitution is a map from the variables of $\mathcal{L}$ to the terms of $\mathcal{L}$. It extends by recursion on $\mathcal{L}$ to a map from $\mathcal{L}$-terms to $\mathcal{L}$-terms.*

Say $R(t_1, t_2)$ iff t_2 is a substitution instance of t_1. (That is to say, iff there is a substitution sending t_1 to t_2.)

The intersection of this preorder with its converse is an equivalence relation. Consider the quotient structure $\langle \mathcal{T}, \leq \rangle$. It has an obvious bottom element, which is the equivalence class containing all the variables of $\mathcal{L}$. Show that (i) $\langle \mathcal{T}, \leq \rangle$ is a lower semilattice, and (ii) if two elements of $\langle \mathcal{T}, \leq \rangle$ have an upper bound, they have a least upper bound.

A declaration of the language of predicate calculus as a recursive datatype was given in section 2.2.2, but no further details were supplied. We will assume that our variables, rather than being x, y, z and so on, are all x's with numerical subscripts. This clearly makes no difference to us, *qua* language users, since it is a trivial relettering, but it does make life a lot easier for us *qua* students of the language. The subscripts are quite important. We call them indices. The purpose of this change in notation is to make visible to the naked eye the fact that we can enumerate the variables: it is much clearer that this is the case if they are written as "x_1, x_2 ..." than if they are written as "$x, y...$"

The universal closure of a formula is the result of prefixing it with enough universal quantifiers to bind all the free variables in it.

Function and predicate letters are not variables, and they cannot be bound with quantifiers. This distinction in the syntax between things that can be bound by quantifiers (the variables) and the things that cannot (the predicate and function letters) sounds like a restriction and therefore a drawback, but it is of fundamental importance and it enables

us to draw useful distinctions. In chapter 1 we were introduced to the
idea of a mathematical object as a set-with-knobs-on. The language of
predicate logic fitted into this picture by assuming that the variables are
intended to range over members of the carrier set, and the predicate and
function letters point to the knobs.

In due course we will explain in detail how this semantics is done,
but we can start with some elementary illustrations. '$(\exists x)(\exists y)(x \neq y)$'
is a formula that is true in those structures with at least two elements.
'$(\exists x)(\exists y)(\exists z)(x \neq y \wedge y \neq z \wedge z \neq x)$' is a sentence true in those
structures with at least three elements. Clearly, for any $n \in \mathbb{N}$ we can
supply a sentence in this style that is true in models with at least n
elements. Trivial though this example is, it serves to make a useful
point: we cannot do this in a way that is *uniform in n*. The temp-
tation to write $(\exists a_1 \ldots a_m)(\forall j, k < m)(k \neq j \rightarrow a_j \neq a_k)$ or even
$(\exists a_1 \ldots a_m)(\bigwedge_{j \neq k < m} a_j \neq a_k)$ must be resisted – in this context at
least.[1] This formula is true in precisely those structures whose car-
rier sets have at least n elements, but it is *not* a formula in the predicate
calculus as the subscripts on the variables are not themselves variables
and cannot be bound. There are plenty of things we can say in predicate logic that cannot be said uniformly, and some of them appear in
the exercises in this chapter.

Less trivial illustrations will concern sets with nontrivial structure.
We have already seen a set of axioms for lattices and a set for boolean
algebras. Structures that can be satisfactorily described by languages
whose variables range only over their carrier set are said to have **first-
order theories**. A property of structures that can be captured by a
formula whose variables range only over elements of the carrier set is said
to be **first-order**. For this reason predicate calculus is sometimes called
first-order logic in contrast to *second-order logic*, where the variables (or
at least some of them) range not over elements of the carrier set but over
subsets of the carrier set. There is also third-order and so on. More of
that later.

There are important connections between logical complexity and computational complexity. Logical complexity theory analyses a property in
terms of how complicated (first-order versus second-order, number of
quantifiers used, etc.) a formula must be to capture it; computational
complexity theory concerns itself with the time it takes to establish

---

[1] This corresponds to an attempt to have variables with internal structure – see the
  discussion on page 91.

whether or not a finite structure has a property in terms of the size of its carrier set. A first-order property can be checked in time bounded by a polynomial in the size of the object being checked for that property, and the degree of the bounding polynomial will be the number of quantifiers in the formula capturing the property. That much is fairly obvious. It would be nice to have a converse to this, namely, a completeness theorem that identifies the syntactic property of being first-order with the semantic property of being checkable in polynomial time. Unfortunately there are counterexamples: there is a polynomial time algorithm to check whether or not a finite group is simple (has no nontrivial normal subgroups), but simplicity is not a first-order property, as we shall see. It is natural to wonder if we can add new constructors to our rectypes of first-order language to obtain an extended concept of first-order language for which converses can be proved. This is an active area of research.

In tackling the following exercises the reader should bear in mind that the way to find a set of first-order axioms for a theory is to remember that first-order means quantifying over elements, not subsets. Identify the property and then the language will write itself.

EXERCISE 53 *Give sets of axioms in suitable first-order languages (to be specified) for the following theories. (These are very roughly in order of difficulty: the first two should be easy and the last two definitely require some thought.)*

  (i) *the theory of integral domains;*
 (ii) *the theory of ordered groups (i.e., groups having a given total order);*
(iii) *the theory of groups of order 60;*
(iv) *the theory of simple groups of order 60;*
 (v) *the theory of algebraically closed fields of characteristic zero;*
(vi) *the theory of partial orders in which every element belongs to a unique maximal antichain;*
(vii) *the theory of commutative local rings (a local ring being a ring with a unique maximal ideal).*

EXERCISE 54 *Which of the following have first-order theories?*

(i) groups all of whose elements are of finite order;
(ii) groups all of whose non identity elements are of infinite order;

(iii) groups with trivial centre;
(iv) groups with an element of infinite order in their centre;
(v) simple groups;
(vi) noetherian rings (rings wherein every ⊆-chain of ideals has a maximal element);
(vii) free groups;
(viii) torsion-free abelian groups (an abelian group is **torsion-free** *if it has no non zero elements of finite order).*

EXERCISE 55 *For $X \subseteq \mathbb{N}$ find a theory T_X that will have a model of size n iff $n \in X$.*

EXERCISE 56 *We say that a formula is* simple existential *when it is of the form $\exists y \phi$, where ϕ is a conjunction of basic formulae (atomic formulae and negations of atomic formulae). Suppose that in a theory T every simple existential formula is equivalent to a quantifier-free formula. Show first that any existential formula $\exists y \psi$ (where ψ is quantifier-free) is equivalent to a quantifier-free formula. Deduce that any formula is equivalent to a quantifier-free formula.*

EXERCISE 57 *Let $\mathcal{C}$ be the first-order language having one binary predicate ϕ_r for each positive rational number r, and let T be the $\mathcal{C}$-theory with axioms (i) $(\forall x)\phi_r(x, x)$ for each $r > 0$, (ii) $(\forall x, y)(\phi_r(x, y) \to \phi_s(y, x))$ for each (r, s) with $r \leq s$ and (iii) $(\forall x, y, z)(\phi_r(x, y) \wedge \phi_s(y, z)) \geq \phi_{r+s}(x, z))$ for each (r, s). Show that every metric space (X, d) becomes a T-model if we interpret '$\phi_r(x, y)$' as '$d(x, y) \leq r$'. Is every T-model obtained from a metric space in this way?*

There is no robust concept of second-order language. There is a difference between (regular) languages that are equal only to the demands of systematically providing names for integers and those languages that can meet the challaenge of providing notations for propositional tautologies, and this difference can be detected by purely syntactic means: the set of strings like '15' and '−123' that name members of Z is a regular language, and the language of propositional logic is not. There is no such *purely syntactic* difference between the predicate calculus and the languages that are intended to be used for doing higher-order mathematics. Here we have another example of the autonomy of syntax. One can set up the language with several distinct suites of variables, so that, for example, lowercase variables can be required to range only over elements of the carrier set, uppercase variables over subsets of the carrier set, and so on. This is common practice, but there is nothing in the langauge that constrains us to consider only those interpretations where the uppercase variables range over *all* subsets of the carrier set. There is nothing to prevent us from using interpretations where, in addition to a carrier set X, one has a designated proper subset of $\mathcal{P}(X)$ as the set over which the uppercase variables range. All one can do is rule that such interpretations are **nonstandard**. So, although there is no concept of a second-order *theory*, there is a concept of a second-order *model* – a model is second order if the designated subset of $\mathcal{P}(X)$ that it includes does indeed contain all subsets of the carrier set, in other words, if it is not nonstandard in that sense.

The **prenex normal form theorem** (PNF) says that every formula of predicate calculus is equivalent to one with all its quantifiers at the beginning, so that every atomic subformula is within the scope of every quantifier.

EXERCISE 58 *Prove the prenex normal form theorem from first principles.*

One of the nice things about the prenex normal form theorem is that it gives us a fairly tidy classification of formulae in terms of complexity. A formula that – once its quantifiers have been pulled to the front – has only universal quantifiers is said to be *universal*; one that similarly has universal quantifiers followed by existential is said to be *universal-existential*. By shoehorning all formulæ into relatively few defined classes with simple definitions like these the prenex normal form theorem provides a framework that makes it natural to state things like: *the class of models of a universal sentence is closed under end-extension,*[1] or *the class of models of a universal-existential sentence is closed under unions of chains.* These things are not that hard to prove, but we would not be naturally motivated to prove them without the PNF.

Analogues of the PNF can be proved for languages intended to be used as higher-order languages, those with several distinct suites of variables intended to range over elements of the carrier set, over subsets of the carrier set, and so on. There is a notation that presupposes these analogues: a formula is Σ_m^n if, once put in normal form, its quantifiers of highest order are of order n and there are m of them. We will not make much of this notation here, but the reader should make note of

---

[1] see page 153.

them. There is even a family of **hierarchy theorems** saying that these classes of formulæ are all distinct.

5.2.0.1 P = NP?

An important class of properties is the class of Σ_1^2 properties: those that can be captured by a formula with one existential second-order quantifier in a suitable second-order language. See Garey and Johnson (1979) for numerous examples. Now, just as it is plausible that a first-order property of a finite structure is checkable in polynomial time (we say it is in P), and that this can be done deterministically, it is plausible that a Σ_1^2 property can be checked nondeterministically in polynomial time ('NP' is short for 'nondeterministic polynomial'). After all, if the property holds of the finite structure, one can verify it by finding a single subset with the right features – and all these features are first-order and can be checked in polynomial time. For a suitably spiced-up first-order language $\mathcal{L}$ this assertion has a converse as well: a property is Σ_1^2 in $\mathcal{L}$ iff it is in NP.

It is not hard to show that there are properties that are captured by Σ_1^2 formulæ that are not captured by (first-order) formulæ of $\mathcal{L}$, and this is what the hierarchy theorems would lead one to expect: Σ_1^2 properties are not in general capturable by first-order formulæ – even spiced-up first-order formulæ. That sounds as if P (the set of problems solvable deterministically in polynomial time) should not be equal to NP. However, in computational complexity theory we are interested only in finite structures, and this raises the question of whether or not Σ_1^2 properties can be captured by first-order formulæ in the sense that for every Σ_1^2 formula there is a first-order formula *with the same finite models.* Thus we can see that $P = NP$ is really a question about how rich the variety of finite structures is: if it is very rich, then there will be a Σ_1^2 formula such that no $\mathcal{L}$-formula is complex enough to have the same finite models and P will not equal NP. If there is not much variety of finite structures then there will be nothing for the extra expressive power of Σ_1^2 formulæ to capture, so P will equal NP.

There are natural and nontrivial examples of distinct theories with the same finite models. A famous one is Wedderburn's theorem, which says that the two theories, of fields and of division rings, despite being different theories, have the same *finite* models. We can show that well-foundedness is not a first-order property (there is no first-order theory whose models are precisely those binary structures $\langle X, R \rangle$ where R is a well-founded relation), but there is a first-order theory ("no loops") whose *finite* models are precisely those binary structures $\langle X, R \rangle$ where R is a well-founded relation and X is finite. Given these examples one is more willing to conjecture that perhaps for every Σ_1^2 property there is a spiced-up first-order property with the same finite models than one would be if one listened merely to the siren voices of the hierarchy theorems telling us that Σ_1^2 is not the same as first-order.

5.3 Formalising predicate logic

As we did in the propositional case in chapter 4, we start with the "many founders, one constructor" point of view, with the intention of abandoning it as promptly here as we did there.

5.3.1 Predicate calculus in the axiomatic style

Add to the three axioms for propositional logic the two new axioms:

$$\forall x A(x) \to A(t); A(t) \to \exists x A(x)$$

and the two new rules of inference:

$$\frac{S \to A(t)}{S \to \forall t A(t)}$$

$$\frac{A(t) \to S}{\exists t A(t) \to S} \qquad \text{'}t\text{' not free in} S.$$

The first of these two rules is often called **universal generalisation**, or **UG** for short. It is a common strategy and deserves a short snappy name. To prove that all F's are G, reason as follows: let x be an F, deduce that x is a G; remark that no assumptions were made about x beyond the fact that it was an F. Conclusion: *all* Fs must therefore be G. The following illustration may help root this rule in your memory.[1]

THEOREM 25 *Every government is unjust.*

Proof: Let G be an arbitrary government. Since G is arbitrary, it is certainly unjust. Hence, by universal generalization, every government is unjust. ∎

---

[1] Thanks to Aldo Antonelli.

5.3.2 Predicate calculus in the natural deduction style

To the natural deduction rules for propositional calculus we add rules for introducing and eliminating the quantifiers:

$$\exists\text{-int} \quad \frac{\vdots \quad A(t)}{(\forall x)(A(x))} \qquad\qquad \exists\text{-elim:} \quad \frac{(\exists x)A(x) \qquad \overset{[A(t)]}{\underset{C}{\vdots}}}{C}$$

$$\forall\text{-int} \quad \frac{\vdots \quad A(t)}{(\forall x)(A(x))} \quad (\text{`}t\text{' not free in the premisses}) \quad \forall\text{-elim} \quad \frac{(\forall x)(A(x))}{A(t)}.$$

However, we will not develop this further but will procede immediately to a sequent treatment.

$\forall$ left:

$$\frac{F(t), \Gamma \vdash \Delta}{(\forall x)(F(x)), \Gamma \vdash \Delta},$$

where 't' is an arbitrary term

$\forall$ right:

$$\frac{\Gamma \vdash \Delta, F(a)}{\Gamma \vdash \Delta, (\forall x)(F(x))},$$

where 'a' is a variable not free in the lower sequent.

$\exists$ left:

$$\frac{F(a), \Gamma \vdash \Delta}{(\exists x)(F(x)), \Gamma \vdash \Delta},$$

where 'a' is a variable not free in the lower sequent.

$\exists$ right:

$$\frac{\Gamma \vdash \Delta, F(t)}{\Gamma \vdash \Delta, (\exists x)(F(x))},$$

where 't' is an arbitrary term.

Reflection on the first footnote on page 71 might help to make sense of the side conditions on the variables in $\exists$ left and $\forall$ right. ("but x was arbitrary, therefore ...").

Notice the similarity between $\vee$-elimination and $\exists$-elimination.

EXERCISE 59 *In this exercise ϕ and ψ are formulæ in which 'x' is not free, while $\phi(x)$ and $\psi(x)$ are formulæ in which 'x' may be free.*

Find proofs of the following sequents:

$\neg\forall x \phi(x) \vdash \exists x \neg\phi(x)$,

$\neg\exists x \phi(x) \vdash \forall x \neg\phi(x)$,

$\phi \wedge \exists x \psi(x) \vdash \exists x(\phi \wedge \psi(x))$,

$\phi \lor \forall x \psi(x) \vdash \forall x (\phi \lor \psi(x)),$

$\phi \to \exists x \psi(x) \vdash \exists x (\phi \to \psi(x)),$

$\phi \to \forall x \psi(x) \vdash \forall x (\phi \to \psi(x)),$

$\exists x \phi(x) \to \psi \vdash \forall x (\phi(x) \to \psi)$

$\forall x \phi(x) \to \psi \vdash \exists x (\phi(x) \to \psi),$

$\exists x \phi(x) \lor \exists x \psi(x) \vdash \exists x (\phi(x) \lor \psi(x)),$

$\forall x \phi(x) \land \forall x \psi(x) \vdash \forall x (\phi(x) \land \psi(x)),$

and deduce the prenex normal form theorem.

5.4 Semantics

We saw in section 2.2.2 how the syntax of predicate calculus (the set of formulæ) can be constructed as a rectype, in a way analogous to the construction of the syntax of propositional logic as a rectype. We also saw how semantics can be given for the syntax of propositional logic by recursion over the datatype of propositional formulæ (see definition 15). The time has now come to do for predicate logic what we did then for propositional logic, namely, provide a recursive semantics. However, predicate logic is powerful and expressive, and it is so similar to natural language in what it appears to be able to do that a few words of warning are in order about what it will *not* achieve.

A lot of semantics for natural languages is not recursive (or "compositional" as the linguists say). There are various ways in which semantics can fail to be compositional. For example, people can use a distinctive vocabulary to announce affiliation to a linguistically defined community – at least in cases where use of that vocabulary was avoidable, because then it represents a choice made by the speaker. People engaged in sports discourse will signal this fact by calling a good player of the game under discussion 'useful'. People who write for the financial press often write 'heading south' for 'decreasing' to signal their membership of this community. Elsewhere, 'represents' for 'is'; 'denotes' for 'is' and 'propose' for 'suggest' mark out the speaker as engaged in scientific discourse, as in the following examples:

Massif-type anorthosites are large igneous complexes of Proterozoic age. They are almost monomineralic, representing [sic] vast accumulations of plagioclase ... the 930-Myr-old Rogaland anorthosite province in Southwest Norway represents [sic] one of the youngest known expressions of such magmatism. (Nature, **405** p.781.)

To divide a number a by a number b means to find, if possible, a number x such that $bx = a$. If such a number exists it is denoted [sic] by a/b. H. Davenport, (1999)

The writers of these examples wished the texts to be read as pieces of scientific discourse and signalled this by a nonstandard use of the flagged word. This part of the author's meaning is not conveyed by building up the meaning of the compound sentence from the meaning of atomic subformulæ by recursion on the structure of the language. Nevertheless, the words still retain some meaning that is revealed compositionally. In contrast, some words used in this way lack compositional semantics altogether: 'elitist' is an example. This word is never used to convey information about the matter under discussion, but only ever to stake a claim by the speaker to be regarded as a person of progressive and egalitarian views. There are certainly other ways in which words in natural language can fail to have entirely compositional semantics, and the transformational grammar of Chomsky and his school is a systematic attempt to capture some of them, but there is no need to explore them here: nonrecursive semantics is very hard to analyse mathematically, and expressions of formal mathematical languages are designed to yield up their meaning without having to be deconstructed in the ways illustrated above.

5.4.1 Truth and satisfaction

In this section we develop the ideas of truth and validity (which we first saw in the case of propositional logic) in the rather more complex setting of predicate logic.

We are going to say what it is for a formula to be **true** in a structure. We will achieve this by doing something rather more general. What we will give is – for each language $\mathcal{L}$ – a definition of what it is for a formula of $\mathcal{L}$ to be true in a structure.

The first thing we need is the concept of a signature from page 48: for a formula ϕ to have a prayer of being true in a structure $\mathfrak{M}$, the signature of the language that ϕ belongs to must be the same as the signature of $\mathfrak{M}$. It simply does not make sense to ask whether or not the transitivity axiom $(\forall xyz)(x < y \wedge y < z. \rightarrow x < z)$ is true in a structure that has no binary relation in it.

First we need to decide what our carrier set is to be. Next we need the concept of an **interpretation**. This is a function assigning to each predicate letter, function letter and constant in the language of ϕ a subset of M^n, or a function $M^k \rightarrow M$, or element of M mutatis mutandis. That is to say, to each syntactic device in the language of ϕ, the interpretation assigns a component of $\mathfrak{M}$ of the appropriate arity.

For example, one can interpret the language of arithmetic by determining that the "domain of discourse" (the carrier set) is to be $\mathbb{N}$, the set of natural numbers, and that the interpretation of the symbol '$\leq$' will be the set of all pairs $\langle x, y \rangle$ of natural numbers where x is less than or equal to y, and so on

If this looks mysterious, it is probably because it really is as banal as you first thought. The problem is that it is *so* banal that one tends to think that something else must have been meant: one *overinterprets*. We remarked earlier that humans are very good at extracting meaning from sentences in nonrecursive ways: they are also quite good at extracting meaning from grammatically defective sentences (my brother-in-law was able to decode the remark he overheard on the bus "My mood-swings keep changing" by means of the fault-tolerant pattern-matching software we all have), but we are not very good at switching these skills off when we do not need them, and they can mislead us – for example by filling in the meaning of '$<$' for us when we should be leaving that task to the official semantics. One is in danger of thinking "What's the point of telling me what '$<$' means? I already know!" The point is that '$<$' *might* have meant something quite different, and the story told here will explain how it might have meant those other things. To use a bit of information technology jargon, one could say that in predicate logic there are no **reserved words**. Well, *almost* anyway. '$=$' is usually taken to be a reserved word and is sometimes called a **logical** predicate letter (though the point is more usually made by referring to all other predicate letters as "nonlogical"). What is meant by this is that '$=$' must always be interpreted by equality. Models not respecting this requirement are said to be **nonstandard** (though there are other ways in which a model might be said to be nonstandard). Of course the quantifiers and connectives are usually taken to be reserved words as well (usually but not always: one needs to reinterpret '$\rightarrow$' when showing that Peirce's law does not follow from K and S: see exercise 48).

We have now equipped the language with an interpretation so we know what the symbols mean, but not what the values of the variables are. In other words, settling on an interpretation has enabled us to reach the position from which we started when doing propositional logic. It's rather like the position we are in when contemplating a computer program but not yet running it. When we run it we have a concept of instantaneous state of the program: these states (snapshots) are allocations of values to the program variables. Let us formalise a concept of state.

A **finite assignment function** is a finite (partial) function from variables in $\mathcal{L}$ to M, the carrier set of $\mathfrak{M}$. These will play a rôle analogous to the rôle of valuations in propositional calculus. I have (see above) carefully arranged that all our variables are orthographically of the form x_i for some index i, so we can think of our assignment function f as being defined either on *variables* or on *indices*, since they are identical up to 1-1 correspondence. It is probably better practice to think of the assignment functions as assigning elements of M to the *indices* and write "$f(i) = \ldots$", since any notation that involved the actual *variables* would invite confusion with the much more familiar "$f(x_i) = \ldots$", where f would have to be a function defined on the things that the variables range over.

Next we define what it is for a partial assignment function to satisfy a sentence p (written "$sat(f, p)$"). We will do this by recursion on the rectype of formulæ, so naturally we define *sat* first of all on atomic sentences.

Notice that in

$$sat(f, x_i = x_j)$$

we have a relation between a function and an expression, not a relation between f and x_i and x_j. That is to say that we wish to **mention** the variables (talk about them) rather than **use** them (to talk about what they point to). This contrast is referred to as the **use-mention distinction.**[1] This is usually made clear by putting quotation marks of some kind round the expressions to make it clear that we are mentioning them but not using them. Now precisely what kind of quotation mark is a good question. Our first clause will be something like

$$sat(f, \,'x_i = x_j') \ \ \text{iff}_{\mathrm{df}} \ f(i) = f(j). \tag{5.1}$$

But how much like? Notice that, as it stands, it contains a name of the expression which follows the next colon: $x_i = x_j$. Once we have put quotation marks round this, the i and j have ceased to behave like variables (they were variables taking indices as values) because quotation is a referentially opaque context.

A context is **referentially opaque** if two names for the same thing cannot be permuted within it while preserving truth. Quotation is referentially opaque because when we substitute one of the two names for Dr. Jekyll/Mr. Hyde for the other in

---

[1] It has been said that the difference between logicians and mathematicians is that logicians understand the use mention distinction.

'Jekyll' has six letters

we obtain the falsehood

'Hyde' has six letters

even though Jekyll and Hyde are the same person. The intuition behind the terminology is that one cannot "see through" the quotation marks to the thing(s) pointed to by the words 'Jekyll' and 'Hyde', so one cannot tell that they are the same. There are other important contexts that are referentially opaque: belief, for example. I might have different beliefs about a single object when it is identified by different names, and these beliefs might conflict.

But we still want the 'i' and 'j' to be variables, because we want the content of clause 5.4.1 to read, in English, something like: "for any variables i and j, we will say that f satisfies the expression whose first and fourth letters are 'x', whose third and fifth are i and j, respectively and whose middle letter is '$=$', iff $f(i) = f(j)$". It is absolutely crucial that in the piece of quoted English text 'x' and '$=$' appear with single quotation marks round them while 'i' and 'j' do not. Formula (5.4.1) does not capture this feature. To correct this Quine invented a new notational device in (1961), which he called "corners" and which are nowadays known as "Quine quotes" (or "quasi-quotes"), which operate as follows: the expression after the next colon:

$$\ulcorner x_i = x_j \urcorner$$

being an occurrence of '$x_i = x_j$' enclosed in Quine quotes is an expression that does not, as it stands, name anything. However, i and j are variables taking integers as values, so that whenever we put constants (numerals) in place of i and j it turns into an expression that will name the result of deleting the quasi-quotes. This could also be put by calling it a variable name.

A good way to think of quasi-quotes is not as a funny kind of quotation mark, for quotation is referentially opaque and quasi-quotation is referentially transparent, but rather as a kind of diacritic, not unlike the LaTeX commands I am using to write this book. Within a body of text enclosed by a pair of quasi-quotes, the symbols '$\wedge$', '$\vee$' and so on, do not have their normal function of composing *expressions* but instead compose *names of expressions*. This also means that Greek letters within the scope of quasi-quotes are not dummies for expressions or abbreviations of expressions but are variables that range over expressions (not sets, or integers). Otherwise, if we think of them as a kind of funny quotation mark, it is a bit disconcerting to find that, as Quine points out, $\ulcorner \mu \urcorner$ is just μ (if μ is an expression with no internal structure). The interested reader is advised to read pages 33-37 of Quine (1961), where this device is introduced.

It might have been easier to have a new suite of operators that combine names of formulæ to get names of new formulæ so that, as it might be, putting '&' between the names of two formulæ gave us a name of the conjunction of the two formulæ named. However, that uses up a whole font of characters, and it is certainly more economical, if not actually clearer, to use corners instead.

Once we have got that straight we can declare the following recursion, where 'α' and 'β' are variables taking expressions as values.

DEFINITION 26 *First the base cases, for atomic fomulæ*

$sat(f,\ulcorner x_i = x_j \urcorner)$ *iff* $f(i) = f(j)$;
$sat(f,\ulcorner x_i \in x_j \urcorner)$ *iff* $f(i) \in f(j)$.

Then the inductive steps
if $sat(f, \alpha)$ *and* $sat(f, \beta)$, *then* $sat(f,\ulcorner \alpha \wedge \beta \urcorner)$;
if $sat(f, \alpha)$ *or* $sat(f, \beta)$, *then* $sat(f,\ulcorner \alpha \vee \beta \urcorner)$;
if for no $g \supseteq f$ *does* $sat(g, \alpha)$ *hold, then* $sat(f,\ulcorner \neg \alpha \urcorner)$;
if there is some $g \supseteq f$ *such that* $sat(g,\ulcorner F(x_i) \urcorner)$, *then*
$sat(f,\ulcorner (\exists x_i)(F(x_i)) \urcorner)$;
if for every $g \supseteq f$ *with* $i \in dom(g), sat(g,\ulcorner F(x_i) \urcorner)$, *then*
$sat(f,\ulcorner (\forall x_i)(F(x_i)) \urcorner)$;

Then we say that ϕ *is* **true in** $\mathfrak{M}$, *written* $\mathfrak{M} \models \phi$ *iff* $sat(\bot, \phi)$, *where* $\bot$ *is the empty partial assignment function. Finally, a formula is* **valid** *iff it is true in every interpretation.*

Remember that this definition was for the toy language of set theory. In other cases the second clause will be replaced by a multiplicity of clauses – one for each predicate.

Beware that $\mathfrak{M} \models T$ and $\Gamma \vdash \Delta$ treat plurals on the right differently. The first means that **everything** on the right is satisfied, the second means only that **something** on the right is satisfied.

DEFINITION 27 *Given a structure* $\mathfrak{M}$, *we write* $Th(\mathfrak{M})$ *for the theory* *of* $\mathfrak{M}$: $\{\phi : \mathfrak{M} \models \phi\}$. *If* $Th(\mathfrak{M}) = Th(\mathfrak{N})$, *we say that* $\mathfrak{M}$ *and* $\mathfrak{N}$ *are* **elementarily equivalent**.

Examples: The reals as an ordered set and the rationals as an ordered set are elementarily equivalent. (Do not try to prove this just yet!) The reals as a field and the rationals as a field are not. Why not? (In the reals every polynomial of odd order has a root!)

If $\mathfrak{M}$ and $\mathfrak{N}$ are elementarily equivalent and $\mathfrak{M}$ is a substructure of $\mathfrak{N}$, we say $\mathfrak{N}$ is an **elementary extension** of $\mathfrak{M}$ if the following extra condition is satisfied: for all expressions ϕ, $\mathfrak{M} \models \phi(\vec{x}) \longleftrightarrow \mathfrak{N} \models \phi(\vec{x})$. (Notice that, because ϕ is allowed to contain free variables, this is a much stronger condition than elementary equivalence.)

Thus the reals as an ordered set are an elementary extension of the rationals as an ordered set. As noted earlier, the reals as a field are not an elementary extension of the rationals as a field.

EXERCISE 60 *The assignment functions we have used have been partial, in contrast to the valuations we used in propositional logic, which were total. How do we have to modify definition 26 if we are to use total assignment functions?*

5.5 Completeness of the predicate calculus

Now that we know what it is for a formula to be true in a model, we have the notion of a valid formula (one that is true in all interpretations) and we can now state and prove the completeness theorem.

THEOREM 28 *A formula in the language of predicate calculus is deducible by the sequent rules iff it is true in all interpretations.*

The strategy is as follows. On being given a formula ϕ, one examines all possible proofs in the hope of finding a proof of ϕ, in a systematic way that ensures that if at the end of time one has failed to find a proof, then the log of one's failed attempts results in a countermodel.

The reader has by now constructed some sequent calculus proofs and seen how one builds a proof in the form of a tree, by backward search. Each node of the tree is decorated by a sequent, and the leaves of the tree are decorated by initial sequents. How does one build the tree? How does one know what to put above a node that is decorated by a sequent that is not an initial sequent? For a start, the root of the tree will be decorated by the sequent $\vdash \phi$, where ϕ is the formula we are trying to prove or find a countermodel for. Any sequent that contains a molecular formula can be the result of applying one of the sequent rules to one or more other sequents. If it contains more than one molecular formula, it can be obtained in more than one way. We guess which way will be most fruitful and put the appropriate sequent(s) above it. For example, the sequent $A \vee B \vdash B \vee C$ can obtained by $\vee$-R from $A \vee B \vdash B, C$ and also by $\vee$-L from the two sequents $A \vdash B \vee C$ and $B \vdash B \vee C$. So if, while we are building a proof tree, we confront a bud decorated with '$A \vee B \vdash B \vee C$', we can either put a bud above it decorated with '$A \vee B \vdash B, C$', or *two* buds, one decorated with '$A \vdash B \vee C$' and the other decorated with '$B \vdash B \vee C$'. How do we choose? It turns out that it does not matter. What we will do is set up in advance a rule that tells us which connectives to attack depending on how far we are from the root of the proof tree under construction. ("If your distance from the root is congruent to 7 mod 13, try to attack conjunctions-on-the-left; if there are no conjunctions-on-the-left, try disjunctions-on-the-right next; if there are no disjunctions-on-the-right try .... On the other hand, if your distance from the root is congruent to 8 mod 13, try disjunctions-on-the-right ... ".)

Not surprisingly, the cases where we do have to be very careful are the four rules for the quantifiers. Let us take $\forall$-L as an illustration. It will become clear that we will have to have an enumeration of the variables of our language, so let us assume one from the outset. We are confronted by a sequent $\Gamma \vdash \Delta$, and we are resolved to attack all the formulæ in Γ that are of the form '$(\forall x_i)\phi$'. We construct a sequent from which $\Gamma \vdash \Delta$ could have been derived by $\forall$-L as follows. For each formula $(\forall x_i)\phi_i(x_i)$ in Γ let 'a_i' be the first variable that we have not already used to attack this formula, but which has appeared in a sequent nearer to the root. Then we add all the formulæ $\phi_i(a_i)$ to Γ to obtain Γ^*. The desired sequent to decorate the new bud is now $\Gamma^* \vdash \Delta$.

The approach to $\exists$-R is exactly the same, so let us investigate $\forall$-R. When we attack $\Gamma \vdash \Delta$ we will add to Δ lots of formulæ $\phi_i(a_i)$ corresponding to the $(\forall x_i)\phi_i(x_i)$ in Δ. This time the a_i we use are those that have *not* appeared in a sequent nearer to the root. Again, $\exists$-L will invite the same approach.

When we have built our tree according to this process, one of two things will happen. Either (i) every branch terminates with an initial sequent, in which case we have a proof, or (ii) there is an infinite path through the tree.[1] In case (ii) we will use the infinite path to construct a countermodel. For one such infinite path, set Γ to be the set of formulæ that appear on the left and Δ to be the set of formulæ that appear on the right. We will build a model in which everything in Γ is true and everything in Δ is false. The carrier set of this model is the set of all the variables. For each n-ary predicate letter R we determine whether or not an n-tuple $\langle x_1 \ldots x_n \rangle$ belongs to the interpretation of R by checking which of Γ or Δ contains $R(x_1 \ldots x_n)$ ■

5.5.1 Applications of completeness

This has been a very cursory teatment of completeness, and several significant details have been omitted. One major topic that has not been covered is the rule of *cut*: from the two sequents $\Gamma \vdash \Delta, \phi$ and $\Gamma', \phi \vdash \Delta'$ infer the sequent $\Gamma \cup \Gamma' \vdash \Delta \cup \Delta'$. If one reads sequents not in the way I have been advocating but instead as saying "if everything on the left is true, then something on the right is true", then this rule is clearly truth-preserving: if one inputs two true sequents, one obtains a true sequent as output. Miniexercise: Check this. (There are two cases, for the two truth-values of ϕ.) Nobody wants to have to use a proof system that includes this rule: backward search using it gives rise to infinitely many possibilities! Thus the fact that it is truth-preserving is very inconvenient: we have to show that everything provable using it is provable without it. The proof of the completeness theorem we sketched proceeds by showing that for any formula there is either a proof of it or a countermodel, and the proof that this process reveals will be cut-free by construction. However, this proof does not give rise to an algorithm that accepts a proof of a formula and outputs a cut-free proof of that formula. The elimination of cuts from proofs is a complex matter for which we do not have space here.

5.5.1.1 Interpolation

There is a precise analogue in predicate calculus of the interpolation lemma for propositional logic, and close attention to the details of the proof of the completeness theorem will enable us to prove it and get bounds on the complexity of the interpolating formula. These bounds are not very good!

---

[1] It might sound as if we need DC for this, but we can pick children of nodes uniformly because the decorations come from a countable set.

The interpolation lemma is probably the most appealing of the consequences of the completeness theorem, since we have very strong intuitions about irrelevant information. Hume's famous dictum that one cannot derive an "ought" from an "is" certainly arises from this intuition. The same intuition is at work in the hostility to the *ex falso sequitur quodlibet* that arises from time to time: if there has to be a connection in meaning between the premisses and the conclusion, then an empty premiss – having no meaning – can presumably never imply anything.

5.5.1.2 Skolemisation

Suppose we have a consistent theory T and that it proves a theorem $(\exists x)\psi(x)$. Then, if $\mathfrak{M}$ is a model of T, we can identify in $\mathfrak{M}$ an element x in $\mathfrak{M}$ such that $\mathfrak{M} \models \psi(x)$. Suppose further that T proves a theorem $(\forall x)(\exists y)\phi(x,y)$. Then, as before, if $\mathfrak{M}$ is a model of T, we can bolt onto $\mathfrak{M}$ a function that, to every x in $\mathfrak{M}$, assigns a y such that $\mathfrak{M} \models \phi(x,y)$. The model that results from bolting this function onto $\mathfrak{M}$ is an *expansion* of $\mathfrak{M}$ (see page 7), and the language for which it is a structure is of course an expansion of $\mathcal{L}(T)$, the language of T. Of course we can do this expansion simultaneously for all ϕ such that T proves $(\forall x)(\exists y)\phi(x,y)$ and adjoin a lot of new function letters to $\mathcal{L}(T)$ in so doing. We can also add, for each ψ such that $T \vdash (\exists x)\psi(x)$, a constant symbol to point to something that is ψ. This process of adding function letters and constant symbols is called **Skolemisation**. The functions denoted in a model by the new function letters are **Skolem functions**. The constants are **Skolem constants**. Whenever we add Skolem functions and constants to a model we can consider the rectype whose founders are the Skolem constants and whose constructors are the Skolem functions. Substructures obtained in this way are natural and important objects.

EXERCISE 61 *(The downward Skolem-Löwenheim theorem) Let T be a consistent theory in a countable language $\mathcal{L}$. Use Skolem functions to prove that T has a countable model.*

5.6 Back and forth

The theory of dense linear order has one primitive nonlogical symbol $\leq$ and the following axioms:

$\forall xyz(x \leq y \to (y \leq z \to (x \leq z))); \forall xz(x \leq y \to y \leq x \to x = y);$

$\forall xy\exists z(x < y \to (x < z \land z < y));$

$\forall x \exists y (y > x);$
$\forall x \exists y (x > y);$
$\forall xy (x \leq y \lor y \leq x).$

THEOREM 29 *All countable dense linear orders without endpoints are isomorphic.*

I shall provide a proof because it is possible to prove the theorem the wrong way.

Proof: Suppose we have two countable dense linear orders without endpoints, $\langle \mathcal{A}, \leq_{\mathcal{A}} \rangle$ and $\langle \mathcal{B}, \leq_{\mathcal{B}} \rangle$. They are both countable, so the elements of $\mathcal{A}$ can be enumerated as $\langle a_i : i \in \mathbb{N} \rangle$ and the elements of $\mathcal{B}$ can be enumerated as $\langle b_i : i \in \mathbb{N} \rangle$.

We start by pairing off a_0 with b_0. Thereafter we procede by induction. At each stage we have paired off some things in $\langle \mathcal{A}, \leq_{\mathcal{A}} \rangle$ with some things in $\langle \mathcal{B}, \leq_{\mathcal{B}} \rangle$. Let us now consider the first thing in $\langle \mathcal{A}, \leq_{\mathcal{A}} \rangle$ not already paired off. (We mean: first in the sense of $\langle a_i : i \in \mathbb{N} \rangle$.) This lies between two things we have already paired, and we must find a mate for it in $\langle \mathcal{B}, \leq_{\mathcal{B}} \rangle$ that lies in the interval between their mates. Since the ordering is dense, this interval is nonempty, and we pick for its mate the first (in the sense of the $\langle b_i : i \in \mathbb{N} \rangle$) in it.

A

B

Fig. 5.1. There is only one countable dense total order without endpoints

Now we consider the first thing (in the sense of the enumeration we have chosen) in $\langle \mathcal{B}, \leq_{\mathcal{B}} \rangle$ not already paired off. This lies between two things we have already paired, and we must find a mate for it in $\langle \mathcal{A}, \leq_{\mathcal{A}} \rangle$ that lies in the interval between their mates. Since the ordering is dense, this interval is nonempty, and we pick for its mate the first (in the sense of the enumeration) in it.

That is the recursive step we use to build the bijection. It goes *back and forth*: $\langle \mathcal{A}, \leq_{\mathcal{A}} \rangle$ to $\langle \mathcal{B}, \leq_{\mathcal{B}} \rangle$ and then $\langle \mathcal{B}, \leq_{\mathcal{B}} \rangle$ to $\langle \mathcal{A}, \leq_{\mathcal{A}} \rangle$. That way we can be sure that by the time we have gone back and forth n times we have used up the first n things in the canonical enumeration of $\langle \mathcal{A}, \leq_{\mathcal{A}} \rangle$

and the first n things in the canonical enumeration of $\langle \mathcal{B}, \leq_\mathcal{B} \rangle$. We will have used n other things as well on each side, but we have no control over how late or early they are in the canonical orderings.

The union of all the finite partial bijections we thus construct is an isomorphism between $\langle \mathcal{A}, \leq_\mathcal{A} \rangle$ and $\langle \mathcal{B}, \leq_\mathcal{B} \rangle$. ∎

Note that this construction shows that the group of order-automorphisms of the rationals acts transitively on unordered n-tuples.

Theorem 29 tells us that the theory of dense linear orders without endpoints is complete. Suppose it were not. Then there would be a formula ϕ that is undecided by it, and by the completeness theorem there would be dense linear orders without endpoints that were ϕ and dense linear orders without endpoints that were not ϕ. But then these dense linear orders would not be elementarily equivalent, and *a fortiori* not isomorphic either.

The study of countable structures that are unique up to isomorphism is a pastime widespread among logicians and has interesting ramifications. A theory is κ-**categorical** iff it has – up to isomorphism – precisely one model of size κ. A *structure* is said to be κ-categorical if its theory is κ-*categorical.*[1] There is a remarkable and deep theorem of Morley that says that a theory that is κ-categorical for even one uncountable κ is κ-categorical for *all* uncountable κ. It is beyond the scope of this book. However, there are a number of natural and important countably categorical theories, and they make good exercises.

5.6.1 Exercises on back-and-forth constructions

EXERCISE 62 *Take two countable dense linear orders without endpoints (for example, two copies of the rationals considered as an ordered set). In both copies paint each point red or blue so that any two red points have a blue point between them and any two blue points have a red point between them. Prove that there is an order-isomorphism between the two copies which respects the colouring.*

The significance of this example is that it is the simplest example of a countably categorical structure whose categoricity has to be proved by a back-and-forth argument and not merely by a "forth" argument.[2] It

---

[1] Readers should be warned that many people confusingly write 'ω-categorical ' when talking about $\aleph_0$-categorical structures like the countable dense linear order without endpoints.

[2] I am endebted to Peter Cameron for pointing this out to me.

seems to be an open question whether or not the countably categorical structures whose uniqueness can be proved by a "forth" construction only is a natural class in any other way.

EXERCISE 63 *Let A_n be the assertion that, if X and Y are disjoint sets of vertices both of cardinality at most n, then there is a vertex x not in $X \cup Y$ joined to every member of X and to no member of Y. Prove that any two countable graphs satisfying A_n for each finite n are isomorphic. (You do not need any results from graph theory to do this.)*

EXERCISE 64 *What is the smallest (nontrivial, i.e., at least two members) graph satisfying A_1? (easy).*

EXERCISE 65 *Any two countable atomless boolean algebras are isomorphic.*

EXERCISE 66 *A countable atomic boolean algebra is **saturated** if every element dominating infinitely many atoms is the sup of two elements dominating two disjoint infinite sets of atoms. Prove that any two countable saturated atomic boolean algebras are isomorphic. (Hint: Show that this condition is equivalent to the requirement that the quotient modulo the ideal of finite elements is atomless.)*

EXERCISE 67 *(The model companion of ZF^-) Let $\gamma(x, y_1 \ldots y_n)$ be a finite conjunction of some of the following atomic formulas and their negations: $x \in x$; $x \in y_i$ $(i \leq n)$; and $y_i \in x$ $(i \leq n)$. We define the theory T as follows. If*

$$\bigwedge_{1 \leq i < j \leq n} y_i \neq y_j \wedge x \neq y_i \wedge \gamma(x, y_1 \ldots y_n)$$

is satisfiable, then

$$(\forall y_1 \ldots y_n)(\exists x)[\bigwedge_{1 \leq i < j \leq n} y_i \neq y_j \rightarrow \bigwedge_{1 \leq i \leq n} x \neq y_i \wedge \gamma(x, y_1 \ldots y_n)]$$

is an axiom of T.
Prove that T is countably categorical .

5.7 Ultraproducts and Łoś's theorem

Let T be a first-order theory. Clearly, if every finitely axiomatised subsystem of T has a model, then every finitely axiomatised subsystem of

T is consistent. This tells us that T itself is consistent (by compactness) and thus that T itself has a model, by the completeness theorem.
Thus we have successfully negotiated our way from the bottom left of
the following diagram to the bottom right:

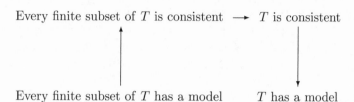

Every finite subset of T is consistent $\longrightarrow$ T is consistent

Every finite subset of T has a model T has a model

We have inferred that T has a model from the news that all its finite
subsets have models, but our proof has involved something very like a
détour through syntax. In the spirit of the interpolation lemma one
might expect that there should be a function that will accept a set of
models and output a model, so that we can feed it models of finite
subsets of T and obtain models of T.

There certainly are constructions that accept sets of models and out-
put (single) models: recall that $\prod_{i \in I} \mathcal{A}_i$ is the direct (sometimes called
Cartesian) product of the $\mathcal{A}_i$.

If $\{\mathcal{A}_i : i \in I\}$ is a family of structures, we define the product

$$\prod_{i \in I} \mathcal{A}_i$$

to be the structure whose carrier set is the set of all functions f defined
on the index set I such that $(\forall i \in I)(f(i) \in A_i)$ and the relations of
the language are interpreted by $R(f, g)$ iff $(\forall i \in I)(R(f(i), g(i)))$. The
$\{\mathcal{A}_i : i \in I\}$ are said to be the **factors** of the product $\prod_{i \in I} \mathcal{A}_i$. For this
operation to make sense it is of course necessary that all the A_i should
have the same signature! (see page 48). Products are nice in various
ways. They preserve horn sentences. What do we mean by "preserve"?

DEFINITION 30 *Let Γ be a class of formulæ. Products preserve Γ if,
whenever $\prod_{i \in I} \mathcal{A}_i$ is a product of a family $\{\mathcal{A}_i : i \in I\}$ and $\phi \in \Gamma$, then
$\prod_{i \in I} \mathcal{A}_i \models \phi$ iff $(\forall i \in I)(\mathcal{A}_i \models \phi)$. In these circumstances we also say
that ϕ is preserved, when $\phi \in \Gamma$.*

By the definition of product, products preserve atomic formulæ.
Clearly they also preserve conjunctions of anything they preserve, and
similarly universal quantifications over things they preserve.

EXERCISE 68 *Verify that products preserve Horn formulæ.*

(This was proved by a man named 'Horn'!) However, they do not always preserve formulæ containing $\vee$ or $\neg$. How so? If ϕ is preserved, then the product will fail to satisfy it if even *one* of the factors does not satisfy it but all the rest do. In these circumstances the product $\models \neg\phi$, but it is not the case that all the factors $\models \neg\phi$. As for $\vee$, if ϕ and ψ are preserved, it can happen that $\phi \vee \psi$ is not, as follows. If half the factors satisfy ϕ and half satisfy ψ, then they all satisfy $\psi \vee \phi$. Now the product will satisfy $\psi \vee \psi$ iff it satisfies one of them. But in order to satisfy one of them, that one must be true at *all* the factors, and by hypothesis it is not. Something similar happens with the existential quantifier.

Given a filter F over the index set, we can define $f \sim_F g$ on elements of the product if $\{i \in I : f(i) = g(i)\} \in F$. This equivalence relation is a congruence relation for all the operations and relations that the product acquires from the factors. At this point it is customary to take a quotient by this congruence relation and call this structure a **reduced product**. This new structure has a different carrier set from the product, but the interpretation of '$=$' in it is indeed equality. It is possible to keep the same carrier set and obtain much of the effect of "reducing" by $\sim_F$ by taking the interpretation of '$=$' in the new structure to be $\sim_F$, but this of course gives a nonstandard model.

Then we *either* take this $\sim_F$ to be the interpretation of '$=$' in the new product we are defining, keeping the elements of the carrier set of the new product the same as the elements of the old, *or* we take the elements of the new structure to be equivalence classes of functions under $\sim$. These we will write $[g]_{\sim_F}$ or $[g]$ if there is no ambiguity. Whichever way you look at it $\sim_F$ is a congruence relation on $\prod_{i \in I} \mathcal{A}_i$.

This new object is denoted by the following expression:

$$\left(\prod_{i \in I} \mathcal{A}_i\right)/F.$$

Similarly, we have to revise our interpretation of atomic formulæ so that

$$\left(\prod_{i \in I} \mathcal{A}_i\right)/F \models R(f,g) \quad \text{iff} \quad \{i : R(f(i), g(i))\} \in F.$$

The reason for proceeding from products to reduced products was to complicate the structure and hope to get more things preserved. In fact, nothing exciting happens (we still have the same trouble with $\vee$ and $\neg$) unless the filter we use is ultra. Then everything comes right.

THEOREM 31 *Łoś's theorem Let $\mathcal{U}$ be an ultrafilter $\subseteq \mathcal{P}(I)$. For all first-order expressions ϕ,*

$$(\prod_{i \in I} \mathcal{A}_i)/\mathcal{U} \models \phi \text{ iff } \{i : \mathcal{A}_i \models \phi\} \in \mathcal{U}.$$

Proof: We do this by structural induction on the rectype of formulæ. For atomic formulæ it is immediate from the definitions.

As we would expect, the only hard work comes with $\neg$ and $\vee$, though $\exists$ merits comment as well.

Disjunction

Suppose we know that $(\prod_{i \in I} \mathcal{A}_i)/\mathcal{U} \models \phi$ iff $\{i : \mathcal{A}_i \models \phi\} \in \mathcal{U}$ and $(\prod_{i \in I} \mathcal{A}_i)/\mathcal{U} \models \psi$ iff $\{i : \mathcal{A}_i \models \psi\} \in \mathcal{U}$. We want to show $(\prod_{i \in I} \mathcal{A}_i)/\mathcal{U} \models (\phi \vee \psi)$ iff $\{i : \mathcal{A}_i \models \phi \vee \psi\} \in \mathcal{U}$.

The steps in the following manipulation will be reversible. Suppose

$$(\prod_{i \in I} \mathcal{A}_i)/\mathcal{U} \models \phi \vee \psi.$$

Then

$$(\prod_{i \in I} \mathcal{A}_i)/\mathcal{U} \models \phi \text{ or } (\prod_{i \in I} \mathcal{A}_i)/\mathcal{U} \models \psi.$$

By induction hypothesis, this is equivalent to

$$\{i : \mathcal{A}_i \models \phi\} \in \mathcal{U} \text{ or } \{i : \mathcal{A}_i \models \psi\} \in \mathcal{U},$$

both of which imply

$$\{i : \mathcal{A}_i \models \phi \vee \psi\} \in \mathcal{U}.$$

$\{i : \mathcal{A}_i \models \phi \vee \psi\}$ is $\{i : \mathcal{A}_i \models \phi\} \cup \{i : \mathcal{A}_i \models \psi\}$. Now we exploit the fact that $\mathcal{U}$ is ultra: for all A and B it contains $A \cup B$ iff it contains at least one of A and B, which enables us to reverse the last implication.

Negation

We assume $(\prod_{i \in I} \mathcal{A}_i)/\mathcal{U} \models \phi$ iff $\{i : \mathcal{A}_i \models \phi\} \in \mathcal{U}$ and wish to infer $(\prod_{i \in I} \mathcal{A}_i)/\mathcal{U} \models \neg\phi$ iff $\{i : \mathcal{A}_i \models \neg\phi\} \in \mathcal{U}$.

Suppose $(\prod_{i \in I} \mathcal{A}_i)/\mathcal{U} \models \neg\phi$. That is to say,

$$(\prod_{i \in I} \mathcal{A}_i)/\mathcal{U} \not\models \phi.$$

By induction hypothesis this is equivalent to

$$\{i : \mathcal{A}_i \models \phi\} \notin \mathcal{U}.$$

But, since $\mathcal{U}$ is ultra, it must contain I' or $I \setminus I'$ for any $I' \subseteq I$, so this last line is equivalent to

$$\{i : \mathcal{A}_i \models \neg\phi\} \in \mathcal{U},$$

as desired.

Existential quantifier

The step for $\exists$ is also nontrivial:

$$(\prod_{i\in I} \mathcal{A}_i)/\mathcal{U} \models \exists x\phi$$

$$\exists f(\prod_{i\in I} \mathcal{A}_i)/\mathcal{U} \models \phi(f)$$

$$\exists f\{i \in I : \mathcal{A}_i \models \phi(f(i))\} \in \mathcal{U},$$

and here we use the axiom of choice to pick a witness at each factor

$$\{i \in I : \mathcal{A}_i \models \exists x\phi(x)\} \in \mathcal{U}.$$

$\blacksquare$

If all the factors are the same, the ultraproduct is called an **ultrapower**, and we write '$A^K/\mathcal{U}$' where K is a set and $\mathcal{U}$ an ultrafilter on K.

THEOREM 32 *The embedding $i = \lambda m_{\mathfrak{M}}.(\lambda f_{\mathfrak{M}^\kappa/\mathcal{U}}.m)$ is elementary.*

Proof: Of course we have to do this by structural induction on formulaæ, but the only hard case is the existential quantifier, and even that is hard in only one direction. After all, if $\mathfrak{M} \models (\exists x)(\phi(x))$, then i of any witness will satisfy ϕ in the ultraproduct. So it will be sufficient to show that, for any $m \in \mathfrak{M}$, if there is an $x \in \mathfrak{M}^\kappa/\mathcal{U}$ such that $\mathfrak{M}^\kappa/\mathcal{U} \models \phi(x, i(m))$, then there is $x \in \mathfrak{M}$ s.t. $\mathfrak{M} \models \phi(x, m)$. Consider such an $x \in \mathfrak{M}^\kappa/\mathcal{U}$. It is the equivalence class $[f]_\mathcal{U}$ of a family of functions f such that $\{\alpha < \kappa : \phi(f(\alpha), m)\} \in \mathcal{U}$. But then this thing in $\mathfrak{M}$ that is $f(\alpha)$ will serve as the witness in $\mathfrak{M}$. $\blacksquare$

Ultraproducts were sold to you a few pages ago as a device that would show that if every finite subset of a theory T has a model, so does T. We had better make this promise good.

Suppose T is a theory (with countably many axioms) such that every finite set of its axioms has a model. Let A_i be a model of the first i axioms of T, and let $\mathcal{U}$ be a nonprincipal ultrafilter on $\mathbb{N}$. Then the ultraproduct $(\prod_{i\in I} \mathcal{A}_i)/\mathcal{U}$ is a model of T (as long as $\mathcal{U}$ is nonprincipal of course! see exercise 38). This has the incredibly useful corollary that

COROLLARY 33 *A formula is equivalent to a first-order formula iff the class of its models is closed under taking ultraproducts.*

EXERCISE 69 *Use ultraproducts to show that well-foundedness is not a first-order property.*

The effect of the ultraproduct construction is to add a lot of things whose presence cannot be detected by finitistic first-order methods. An important effect of this is a direct proof of

THEOREM 34 *(Upward Skolem-Löwenheim theorem) Every consistent theory with an infinite model has arbitrarily large models.*

Proof: We can prove this by appealing to the completeness theorem. If T is a consistent theory with an infinite model, add to the language of $\{L(T)$ as many constant symbols 'a_i', 'a_j' as you please, and add to T axioms '$a_i \neq a_j$' saying that all these constants are distinct. The compactness theorem for predicate logic ensures that this new theory is consistent, and the completeness theorem for predicate logic proves that it has a model.

However, we can instead prove it directly using ultraproducts. Let $\mathfrak{M}$ be a model of T, K an index set of cardinality κ a cardinal as large as you please, and $\mathcal{U}$ an ultrafilter on K. $\mathfrak{M}^K/\mathcal{U}$ is then elementarily equivalent to $\mathfrak{M}$ and is large. How large? Well, an element of $\mathfrak{M}^K/\mathcal{U}$ is an equivalence class of functions from K to M. To show that there are many equivalence classes we must show that there are large families of functions that pairwise disagree on a set in $\mathcal{U}$. But for each address in K we can pick any one of $|M|$ distinct things, and this gives us κ independent choices of things from M, so there are $|M|^\kappa$ functions that pairwise differ everywhere. This size is certainly at least as big as κ. ∎

Theorem 34 can be strengthened to assert that every theory with an infinite model has models of all larger sizes.

5.7.1 Further applications of ultraproducts

Keisler's ultrapower lemma connects the logical concept of elementary equivalence with the algebraic concept of isomorphism. It says that, if two structures $\mathfrak{M}$ and $\mathfrak{N}$ are elementarily equivalent, then they have ultrapowers $\mathfrak{M}'$ and $\mathfrak{N}'$ that are isomorphic. The converse is easy, but the hard direction is beyond the scope of this book.

5.7.1.1 Nonstandard models of arithmtic

In an ultrapower of the reals one finds elements like the equivalence class of the function $\lambda n.(1/n)$. This is clearly an infinitesimal: it is everywhere bigger than 0 and, for each n, eventually less than $1/n$. This enables us to do something the eighteenth century wanted to do but couldn't, namely, do differential and integral calculus using infinitesimals, and do it rigorously. Presenting analysis in this way has not caught on, but it well might yet. See Keisler (1976a, 1976b).

5.8 Exercises on compactness and ultraproducts

EXERCISE 70 *Show that the theory of connected graphs is not first-order.*

EXERCISE 71 *Write down first-order axioms for the theory of fields. Show that if a first-order statement is true in all fields of characteristic 0, then it is true in all fields of sufficiently large characteristic.*

EXERCISE 72 *Write down the axioms for an ordered field (essentially the axioms for the reals without the crucial completeness axiom). An ordered field is* archimedean *just when for every $x > 0$ there is $n \in \mathbb{N}$ with $x + x + x + x + x \ldots (n$ times$) > 1$. Show that there exist non-archimedean ordered fields.*

EXERCISE 73 *Show that if $\mathcal{U} \subseteq \mathcal{P}(I)$ is the principal ultrafilter generated by j, then $(\prod_{i \in I} \mathcal{A}_i)/\mathcal{U} \simeq \mathcal{A}_j$.*

EXERCISE 74 *This question is designed to take about 45 minutes.*

A **pedigree** *is a set P with two unary total functions f amd m defined on it, with disjoint ranges. ($m(x)$ is x's mother and $f(x)$ is x's father.)*

(i) Set up a first-order language $\mathcal{L}$ for pedigrees and provide axioms for a theory T_1 of pedigrees.

A pedigree may be **circle-free***: in a realistic pedigree no one is their own ancestor! Realistic pedigrees are also* **locally finite***: no one is the father or mother of infinitely many things.*

(ii) One of these two new properties is first-order and the other is not. Give axioms for a theory T_2 of the one that is first-order and an explanation of why the other one is not.

A **fitness function** *is a map v from P to the reals satisfying $v(x) = (1/2) \cdot \Sigma_{f(y)=x} v(y)$ or $v(x) = (1/2) \cdot \Sigma_{m(y)=x} v(y)$ (depending on whether x is a mother or a father).*

(iii) Find a sufficient condition for a pedigree to have a nontrivial fitness function and a sufficient condition for it to have no nontrivial fitness funtion.

(iv) Extend your language $\mathcal{L}$ to include syntax for v. In your new language provide axioms for a new theory T_3 that is to be a conservative extension of T_1 and whose locally finite models are precisely the locally finite pedigrees with a nontrivial fitness function.

There is an obvious concept of generation for a pedigree.

(v) Expand $\mathcal{L}$ by adding new predicate(s) and give a first-order theory in this new language for pedigrees that have well-defined generations. Give first-order axioms in $\mathcal{L}$ itself for a theory of pedigrees that have well-defined generations.

6

Computable functions

Hilbert's 1900 address set a number of tasks whose successful completion would inevitably involve more formalisation. It seems fairly clear that this was deliberate: Hilbert certainly believed that if formalisation was pursued thoroughly and done properly, then all the contradictions that were crawling out of the woodwork at that time could be dealt with once and for all.

One of the tasks was to find a method for solving all diophantine equations. What does this mean exactly? For example, it is easy to check that for any two naturals a and b

$$(a^2 - b^2)^2 + (2ab)^2 = (a^2 + b^2)^2,$$

and so there are infinitely many integer solutions to $x^2 + y^2 = z^2$ (the pythagorean equation). Indeed we can even show that every solution to the pythagorean equation (at least every solution where x, y and z have no common factor) arises in this way:

Notice that if $x^2 + y^2 = z^2$, then z is odd and precisely one of x and y is even. (We are assuming no common factors!) Let us take y to be the even one and x the odd one.

Evidently $x^2 = (z - y)(z + y)$m, and let d be the highest common factor of $z - y$ and $z + y$. Then there are coprime a and b satisfying $z + y = ad$ and $z - y = bd$. So $x^2 = abd^2$. This can happen only if a and b are perfect squares, say u^2 and v^2, respectively. So $x = uvd$.

This gives us $z = \frac{u^2 + v^2}{2} \cdot d$ and $y = \frac{u^2 - v^2}{2} \cdot d$ and in fact d turns out to be 2.

Hilbert's question – and it is a natural one – was, can we clean up all diophantine equations in the way we have just cleaned up this one?

If there is a general method for solving diophantine equations, then we have the possibility of finding it. If we find it, we exhibit it, and we are done. To be slightly more specific, we have a proof that says "Let E be a diophantine equation, then ...", using the rule of universal generalisation (UG).

On the other hand, if there is no such general method, what are we to do? Merely gesticulating despairingly in front of hard cases will not persuade anyone that those cases cannot be solved. We would have to say something like: let $\mathfrak{A}$ be an arbitrary algorithm; then we will show that there is a diophantine equation that $\mathfrak{A}$ does not solve. But clearly, in order to do this, we must have a formal concept of an algorithm. Hilbert's challenge was to find one.

There are various formal versions of computation. We saw finite-state machines earlier, and we saw how the set of strings recognised by a machine gives rise to a concept of computable set. However, we also saw a fatal drawback to any analysis of computable set in terms of finite-state machines: the matching bracket language is not recognised by any finite-state machine but is obviously computable in some sense. The problem arises because each finite-state machine has a number of states (or amount of memory, to put it another way) that is fixed permanently in advance. The most general kind of computation that we can imagine that we would consider to be computation is deterministic, finite in time and memory but unbounded: no predetermined limit on the amount of time or memory used. There have been various attempts to capture this idea in machinery rigorous enough for one to prove facts about it. The (historically) first of the most general versions is Turing machines. There is also representability by λ-terms. This is a rich and fascinating branch of logic that we cannot treat here: there is too much of it, and, in any case there are many elegant treatments of it in print already. Another attempt is μ-recursion, which we will do in detail below. The other approach we explore in some detail is an analysis in terms of register machines. They do not have the historical *cachet* of Turing machines but are slightly easier to exploit, since they look more like modern computers.

What became clear about 70 years ago is that all attempts to formalise the maximal idea of a computable function result in the same class of functions. This gives rise to **Church's thesis**. Although not normally presented as such, Church's thesis is really just a claim that this endeavour to illuminate – by formalisation – our intuitive idea of a computable function has now been completed: we will never need another notion of computable.

Philosophically inclined readers may wish to reflect on the curious fact that Church's thesis is a metamathematical allegation of which no formal proof can be given. As soon as we formalise "All attempts to formalise our informal notion of finite-but-unbounded computation result in the same formal notion", the reference to informal computation becomes a reference to a formal notion and the sense is lost.

How can we be so confident? Well, we have a completeness theorem. All completeness theorems have two legs: a semantic concept and a syntactic concept. The semantic concept in this case is Turing-computable or register machine-computable. The syntactic concept is a bit harder. The first attempt at it is **primitive recursive**; we will discover the correct syntactical concept by examining what goes wrong with primitive recursive functions.

6.1 Primitive recursive functions

DEFINITION **35** *The rectype of* **primitive recursive functions** *is the $\subseteq$-least class of functions containing the* **initial** *functions, which are*

the **successor** *function:* $\lambda n.n + 1$, *written* S;
the **projection** *functions:* $proj_n^m$ *accepts an m-tuple and returns its nth entry;*
the **constant** *functions: for each k and n there is a function that accepts a k-tuple and returns the output n*

and closed under (i) **composition**

The composition operation under which the rectype of primitive recursive function declarations is closed is slightly more complicated than the familiar composition of functions of one argument. It is like composition of terms (see page 47, clause 5). It is fiddly but obvious. It is worth noting though that, if $\lambda xy.f(x, y)$ is a primitive recursive function of two variables, then $\lambda x.f(x, x)$ is a primitive recursive function of one variable.

and (ii) **primitive recursion***:*

$$f(\vec{x}, 0) := g(\vec{x}); \qquad f(\vec{x}, y + 1) := h(\vec{x}, y, f(\vec{x}, y)). \qquad (6.1)$$

We say f is declared by **primitive recursion** *over g and h. Notice that, although there is no limit on the number of variables we can compute with, we only recurse on one.*

On the face of it this declaration looks very restrictive: "only allowed one call", but it turns out to be surprisingly fertile.

Strictly what we have here is a rectype of *function declarations* rather than *functions*. We will think of function declarations as a kind of function-in-intension and will consider a function(-in-extension) to be primitive recursive if it has a primitive recursive declaration (as a functionin-intension).

Note at the outset that this datatype of function declarations is countably presented (see section 2.1.6) and so has only countably many elements.

The basic functions are in some obscure but uncontroversial sense computable; clearly the composition of two computable functions is computable, and if g and h are in some sense computable, then f declared over them by primitive recursion is going to be computable in the same sense. That is why this definition is *prima facie* at least a halfway sensible stab at a definition of computable function.

We adopt the habit of bundling together all the snail variables (those you just carry around and do not recurse on) in the style $\vec{x}$.

We need 'y' in the right-hand side of the second clause of definition 6.1 because otherwise, if it should ever happen that there are n and k such that $f(\vec{x}, n) = f(\vec{x}, k)$, we will have $f(\vec{x}, n+1) = f(\vec{x}, k+1)$ and $\lambda n.f(\vec{x}, n)$ will be periodic.

Here are some declarations:

(i) Predecessor: $P(0) := 0$; $P(S(x)) := x$.

(ii) Bounded subtraction: $x \doteq 0 := x$; $x \doteq S(y) := P(x \doteq y)$.

(iii) Addition: $x + 0 := x$; $x + S(y) := S(x+y)$.

(iv) Multiplication: $x \cdot 0 := 0$; $x \cdot (S(y)) := (x \cdot y) + x$.

EXERCISE 75 *Show that if f is primitive recursive, so are*
(i) the function that returns the sum of the first n values of f;
(ii) the function that returns the product of the first n values of f.

6.1.1 Primitive recursive predicates and relations

A predicate $R(\vec{x})$ is a **primitive recursive predicate** (or relation) iff there is a primitive recursive function r such that $r(\vec{x}) = 0 \longleftrightarrow R(\vec{x})$. We can take 1 to be **true** and 0 to be **false**, or vice versa, or 0 to be **true** and all other values to be **false** – it does not matter as long as one is consistent. In what follows **true** is 1.

EXERCISE 76 *Show that $\leq$ is a primitive recursive relation.*

If R and S are primitive recursive predicates represented by r and s, then

$\neg R$ is represented by $1 \mathbin{\dot-} r$;

$R \wedge S$ is represented by $r \cdot s$;

$R \vee S$ is represented by $r + s \mathbin{\dot-} (r \cdot s)$;

$(\exists x \leq z)(R(x, \vec{y}))$ is represented by

$$\prod_{0 \leq x \leq z} r(x, \vec{y}).$$

We can capture bounded universal quantification by exploiting duality of the quantifiers.

The set of primitive recursive functions is also closed under `if then else`, in the sense that if r is a primitive recursive predicate, then `if` R `then` x `else` y is also primitive recursive. Here's why. Declare

$$\texttt{if-then-else}(x, y, 0) := x; \ \texttt{if-then-else}(x, y, S(n)) := y.$$

`if-then-else` is evidently primitive recursive, and it is mechanical to check that

$$\texttt{if-then-else}(proj(r, x, y)_2^3, proj(r, x, y)_3^3, proj(r, x, y)_1^3)$$

evaluates to x if $r = 1$ and to y if $r = 0$.

Putting this together with the fact that bounded quantification is primitive recursive tells us that functions declared in the style

$$\texttt{if} \ (\exists x < y)R(x) \ \texttt{then} \ f(x) \ \texttt{else} \ g(x).$$

are primitive recursive, as long as R, f and g are. This is **bounded search**. Hofstader (1979) memorably calls this "BLOOP".

The order relation on the rectype $\mathbb{N}$ is its engendering relation, the transitive closure of the constructor S. This motivates very sweetly the restricted quantifiers that we have just considered. Before we leave this digression about bounded quantification we must make the point that bounded quantifiers give us an analogue of the prenex normal form theorem.

EXERCISE 77 *Show that any expression in the language of arithmetic is equivalent to one in which all the bounded quantifiers are within the scope of all the unbounded quantifiers. (You may take pairing and un-pairing functions to be primitive functions.)*

Hint: The only hard part is dealing with $(\forall x < y)(\exists k)$.

In the following questions you may assume that `pair` represents a primitive recursive bijection $\mathbb{N}^2 \to \mathbb{N}$. The following is a standard example:

$$\texttt{pair}(x, y) = (^1/_2) \cdot (x^2 + y^2 + 3x + y + 2xy)$$

and `fst` and `snd` are the corresponding primitive recursive unpairing functions (so that $\texttt{fst}(\texttt{pair}(m, n)) = m$, $\texttt{snd}(\texttt{pair}(m, n)) = n$ and $\texttt{pair}(\texttt{fst}(r), \texttt{snd}(r)) = r$).

EXERCISE 78

(i) *The declaration:*

     $\texttt{Fib}(n + 2) := \texttt{Fib}(n + 1) + \texttt{Fib}(n); \texttt{Fib}(1) := 1; \texttt{Fib}(0) := 1$

     is not primitive recursive. Find a declaration of this function that is primitive recursive.

(ii) *The iterate* $\mathrm{It}(f)$ *of* f *is defined by:* $\mathrm{It}(f)(m, n) = f^m(n)$. *Notice that, even if* f *is a primitive recursive function of one argument, this function of two arguments is not* prima facie *primitive recursive. Show that it is primitive recursive nevertheless.*

     Take $\mathcal{I}$ *to be the inductively defined class of functions containing the successor function* $S(n) = n + 1$, *the functions* `pair`, `fst`, `snd` *and closed under composition and iteration. Show that if* $a \in \mathbb{N}$ *and* $G(x, y)$ *is in* $\mathcal{I}$ *and* $H(x)$ *is defined by* $H(0) = a$, $H(n + 1) = G(H(n), n)$, *then* $H(x)$ *is in* $\mathcal{I}$. *[Hint: Consider* $\texttt{pair}(H(y), y).]$

EXERCISE 79 *Show that all primitive recursive functions are total by induction on the rectype. The induction step for primitive recursion uses induction over* $\mathbb{N}$.

This means that functions like that which returns n when given $2n$ and fails on odd numbers is not primitive recursive. Nevertheless, you will often hear people say – as I say to you now – that you would be extremely unlucky to encounter computable functions that are not primitive recursive unless you are a logician and go out of your way to look for trouble. The resolution of this apparent contradiction is that the function $\lambda n.(\texttt{if } n = 2k \texttt{ then } k \texttt{ else fail})$ is in some sense *coded* by the primitive recursive function that sends $2n + 1$ to 0 (meaning `fail`) and sends $2n$ to $n + 1$ (meaning n), and this function *is* primitive recursive.

6.2 μ-recursion

Does the rectype of primitive recursive functions exhaust the class of (total) functions that reasonable people would consider computable?

DEFINITION 36 *Ackermann function:*

$$A(0,y) := y+1; \quad A(x+1,0) := A(x,1); \quad A(x+1,y+1) := A(x, A(x+1,y)).$$

DEFINITION 37 f **dominates** g *if, for all sufficiently large n, $f(n) > g(n)$.*

EXERCISE 80 *For every primitive recursive function $f(\vec{x}, n)$ there is a constant c_f such that*

$$(\forall n \forall \vec{x})(f(\vec{x}, n) < A(c_f, max(n, \vec{x}))).$$

(In slang, every primitive recursive function is in $O(Ackermann)$.)

(Hint: Use induction on the rectype of primitive recursive functions.)

COROLLARY 38 *The Ackermann function is therefore not primitive recursive.*

But it is still total!

REMARK 39 $A(n, m)$ *is defined for all $n, m \in \mathbb{N}$.*

Proof: We need to recall that the lexicographic product $\mathbb{N} \times \mathbb{N}$ is a well-order. This means that we can do well-founded induction on it. Let $\langle x, y \rangle$ be minimal in the lexicographic order of $\mathbb{N} \times \mathbb{N}$ such that $A(x, y)$ is undefined. It does not take long to check that y and x must both be nonzero. But in these circumstances $A(x, y) := A(x-1, A(x, y-1))$. Now the pair $\langle x, y-1 \rangle$ is below $\langle x, y \rangle$ in the lexicographic order of $\mathbb{N} \times \mathbb{N}$, so $A(x, y-1)$ is defined, so we can use the fact that $\langle x-1, A(x, y-1) \rangle$ is below $\langle x, y \rangle$ in the lexicographic order of $\mathbb{N} \times \mathbb{N}$ to infer that $A(x-1, A(x, y-1))$ must be defined (since $\langle x, y \rangle$ was minimal in the lexicographic order of $\mathbb{N} \times \mathbb{N}$ such that $A(y, x)$ is undefined!). So $A(x, y)$ (which is $A(x-1, A(x, y-1))$) is defined after all! Contradiction. ∎

EXERCISE 81 *For* **enthusiasts** *only![1] When you are satisfied with your answer to exercise 80 – and you should be! – try what follows:*

(i) *Write out a definition of a constructor of* **double recursion** *so that you now have a rectype of doubly recursive functions. (Do not worry unduly about how comprehensive your definition is.)*

(ii) *What would a ternary Ackermann function be? Prove that the ternary Ackermann function you have defined dominates all doubly recursive functions in the manner of your proof of exercise 80.*

The Ackermann function involves recursion on two variables in a way that cannot be disentangled. The point of exercise 81 is that there is also treble recursion and so on. A function is *n*-**recursive** if it is declared by a recursion involving n entangled variables. Exercise 81 invites you to prove analogues for each n of the facts we have proved about the Ackermann function: namely, for every n there are functions that are n recursive but not $n-1$-recursive, and one can prove their totality by a well-founded induction over the lexicographic product $\mathbb{N}^n$. Is every total computable function n-recursive for some n? Sadly, no, but I shall not give a proof. It turns out that the correct response to the news brought by the Ackermann function that not every total computable function is primitive recursive is not to pursue 2-recursive, 3-recursive and so on but rather to abandon altogether the idea that computable functions have to be total in order to be computable. For a sensible general theory we need to consider **partial** functions.[2] This is because we want unbounded search[3] to be allowed. The new gadget we need is μ-recursion, which corresponds to unbounded search. This is a sensible new constructor to reach for because any strategy for computing g will give rise to a strategy for computing g^{-1}: simply try g with successively increasing inputs starting at 0 and continue until you get the answer you want – if you ever do. The point is that, if we have a deterministic procedure for getting values of g, we will have a deterministic procedure for getting values of g^{-1}. That is to say, it appears that the class of functions that

---

[1] Then why put it in!? Because it makes a point I shall need to allude to later: read it but do not do it.

[2] On page 128 we encountered a naturally occurring computable partial function that was not *really* partial because there was a computable total function that in some sense encoded the same information. When I write that we must embrace partial functions I mean we must embrace even those partial functions that cannot be coded as total function in the way division by 2 can.

[3] Gödel Escher Bach fans might be helped by a reminder that Hofstader calls unbounded search FLOOP (as opposed to BLOOP, which is bounded search).

are plausibly computable (in an intuitive sense of 'computable') is closed under inverse.

So we augment the constructors of the rectype of primitive recursive functions by allowing ourselves to declare f by $f(n, \vec{x}) := (\mu y)(g(y, \vec{x}) = n)$, once given g. Then $\mu y.\Phi$ is the least y such that Φ (if there is one) and is undefined otherwise.

Notice that, even with this new constructor, the rectype of μ-recursive functions is still countably presented.

But there is a catch to this. The unbounded search constructor preserves computability as long as its argument is a total function, but the inverse function that it gives us is not guaranteed to be total itself! Think about inverting $\lambda n.2n$. The result is a function that divides even numbers by 2 and fails on odd numbers. No problem there. For the moment let f be the function that divides even numbers by 2 and fails on odd numbers. The problem arises if we try to invert f: how do we ever discover what $f^{-1}(3)$ is? It ought to be 6 of course, but if we approach it by computing $f(0)$, $f(1)$ and so on, we get stuck because the endeavour to compute $f(1)$ launches us on a wild goose chase. We could guess that the way to compute $f^{-1}(3)$ is to try computing $f(6)$, but we do not want to even think about nondeterminism, because this severs our chain to the anchor of tangibility that was the motivation for thinking about computability in the first place.

The upshot is that we cannot rely on being able to iterate inversion, so we cannot just close the set of primitive recursive functions under the old constructors and this new one and expect to get a sensible answer. As the $\lambda n.2n$ example shows, FLOOP might output a function that you cannot then FLOOP. Nor can we escape by doctoring the datatype declaration so that we are allowed to apply inversion only to functions satisfying conditions that – like totality – are ascertainable solely at run-time. That would not be sensible.[1]

---

[1] It is true that one can obtain a declaration of the μ-recursive functions as a rectype by simply adding to the constructors for the primitive recursive functions the declaration:

If $\phi(\vec{x}, y)$ is a total μ-recursive predicate, then $f(\vec{x}) := (\mu y)(\phi(\vec{x}, y) = 0)$ is a μ-recursive function.

and some writers do this, but this is philosophically distasteful for the reasons given: it makes for a less abstract definition.

Fortunately it will turn out that any function that we could define by more than one inversion can always be defined by only one.[1] I am going to leave the *precise* definition of μ-recursive up in the air for the moment. We will discover what it is by attempting to prove the theorem that a function is μ-recursive iff it is computable by a machine

At first blush it seems odd to formalise computability in such a way that a function can be computable but undefined, but this liberalisation is the key that unlocks computation theory. Perhaps on reflection it isn't so odd after all: all of us who have ever written any code at all know perfectly well that the everywhere undefined function is computable – since we have all inadvertently written code that computes it!

Specifically, this enables us to connect syntactic concepts of computability, namely, function declarations, to semantic concepts, namely, computability by machines, to which we now turn.

6.3 Machines

I mentioned earlier that it does not matter what kind of architecture our machines have as long as they have unbounded memory and can run indefinitely. The paradigm we use for the sake of illustration is register machines.

A register machine has

(i) finitely many registers $R_1 \ldots R_n$ each of which holds a natural number; and

(ii) A **program** that is a finite list of **instructions** each of which consists of a **label** and a **body**. Labels are natural numbers, and a body has one of the three forms:

  (i) $R^+ \to L$: add 1 to contents of register R and jump to instruction with label L.
  (ii) $R^- \to L', L''$: if contents of R is nonzero, subtract 1 from it and jump to the instruction with label L' otherwise. jump to the instruction with label L''.
  (iii) HALT!

The **output** of the register machine is the contents of register 1 (say) when the machine executes a HALT command. Notice that we don't really specify the number of registers by stipulation but only indirectly by mentioning registers in the instructions in the program. If the program

---

[1] Unfortunately this is not proved by exhibiting an algorithm for eliminating extra inversions: it's less direct than that.

has only ten lines, it cannot mention more than ten registers, and so the machine can be taken to have only ten registers.

We say that a register machine $\mathfrak{M}$ **computes** a function f iff, for all $n \in \mathbb{N}$, $f(n)$ is defined iff whenever we run $\mathfrak{M}$ starting with n in register 1 it halts with $f(n)$ in register 1 and does not halt otherwise.

It is very important that the register machines can be effectively enumerated, but deeply unimportant how we do it, though one can collect a few hints.[1]

Recall the discussion on page 40. The prime powers trick lets us code lists of numbers as numbers. If we do this, the usual list-processing functions `head`, `tail` and `cons` will be primitive recursive. Although it is simultaneously very important that the register machines can be effectively enumerated yet deeply unimportant how we do it, there is one fact about how we do it that we will need, and that is that the map from numbers to machines should be computable in some sense. We can describe a machine completely in a specification language of some kind, because a machine is after all a finite object, and it will have a finite description, and we can have a standardised uniform way of presenting these descriptions. The specification language can be written in an alphabet with perhaps 50 characters (alphanumerics and punctuation), so if we identify a machine with its description in the language, it can be thought of as a numeral to base 50. This numeral will not be a mere *name* of the machine, but an actual *description* of it.

$\mathbb{N}$ is a rectype, and so is the set of machine descriptions in the specification language. The gnumbering function given is nice in the sense that it is a rectype homomorphism.

If a formula is a list of symbols, we can define a Gödel enumeration of formulæ by list-recursion as shown in the following ML pseudocode. The

---

[1] Indeed it is deeply important that it is unimportant, for this is another *invariance* point:

"That's very important," the King said, turning to the jury. They were just beginning to write this down on their slates, when the White Rabbit interrupted: "*Un*important, your Majesty means, of course," he said in a very respectful tone, but frowning and making faces at him as he spoke.

"*Un*important, of course, I meant," the King hastily said, and went on to himself in an undertone, "important – unimportant – unimportant – important – " as if he were trying which word sounded best.

Some of the jury wrote it down "important," and some "unimportant". Alice could see this, as she was near enough to look over their slates; "but it does not matter a bit," she thought to herself.

gnumber of a formula is a number to base 256 (because ASCII codes are numbers below 256!).

```
gnumber h::[ ] = ASCII of h
   |      h:: t  = 256*gnumber(t) + gnumber(h);
```

From now on we are going to assume we have fixed an enumeration of register machines in this style, so that the mth machine is the machine with gnumber m. There is a convention of writing "$\phi_e(n) \downarrow = k$" to mean that the eth machine halts with input n and outputs k. "$\phi_e(n) \uparrow$" means that the eth machine does not halt with input n. In these circumstances we say $\phi_e(n)$ **diverges**.

6.3.1 The μ-recursive functions are precisely those computed by register machines

An essential gadget is

DEFINITION 40 *(Kleene's T function) Input m and i and t, output a list of t states of the mth machine started with input i, one for each time $t' < t$. (The state of a register machine is the tuple of contents of the registers and the current instruction.)*

The output, $T(m, i, t)$, of Kleene's T-function is commonly called a **complete course of computation**. It is entirely plausible that T is computable since, as long as the gnumbering is sensible in the sense that the gnumber of a machine is a *description* of it, and the machines have standard architecture, on being given a gnumber m one can go away and build the machine described by m and then feed it input i and observe it for t steps. It is a lot less obvious that T is primitive recursive, but it is. The proof would be extremely laborious, but it relies merely on checking that all the functions involved in encoding and decoding are primitive recursive. This is plausible because the machines have finite descriptions and are deterministic. Not only are they deterministic, but the answer to the question, "What state will it go to next?" can be found by looking merely at the machine and its present state, without consulting the positions of the planets or anything else that – however determinate – is not internal to the machine.

This shows that

THEOREM 41 *The function computed by the mth machine is μ-recursive.*

In other words, the machine with gnumber m computes the μ-recursive function: $\lambda i.$ the least k such that m started with i halts with output k.

Now for the converse.

THEOREM 42 *Every μ-recursive function can be computed by a register machine.*

Sketch of proof: Consider the rectype of functions built up from the initial functions (as in the declaration of primitive recursive functions) by means of composition, primitive recursion and μ-recursion. This class contains all sorts of functions that are undefined in nasty ways because it allows us to invert the results of inversions, and the result of inverting a function might not be total – as we have seen. Nevertheless, we can prove by induction on this datatype that for every declared function in it there is a register machine that computes it. That is, in the sense that whenever these declarations do not fall foul of common sense by attempting to invert functions that are not total, the machine that we build does indeed compute the function.

The details of how to glue together register machines for computing f and g into one that computes $f \circ g$ will be omitted, as will the details of how to compose register machines to cope with the primitive recursion constructor, and how to front-end something onto a register machine that computes $f(x, y, \vec{z})$ to get something that computes $\mu x.(f(x, y, \vec{z}) = k)$. ∎

This completes the proof of the completeness theorem for computable functions.

6.3.2 A universal register machine

Kleene's T-function is primitive recursive, so there is a machine that computes it. Any such machine can be tweaked into a **universal** or all-purpose machine: one that can simulate all others.

We need several auxilliary functions on memory-dumps:

(i) `current_instruction`(d) and `register_0`(d), which return respectively the current instruction and the contents of register 0;

(ii) `last` returns the last element of a list.

It is mechanical to check they are all primitive recursive. Once we have got those, we can build a machine that, on being given m and i, outputs:

`register_0(last(`$T(m, i, (\mu t)($`current_instruction(last(`$T(m, i, t) =$
`HALT))))))`

which is what the mth machine does on being given i.

One of the intentions behind the invention of computable functions was to capture the idea of a decidable set. One exploits it in some manner along the following lines. A set is decidable iff it is the range of a computable function. It turns out that that does not straightforwardly give us what we want. Suppose we want to know whether or not n is a member of a putatively decidable set, presented as f"$\mathbb{N}$, for some computable function f. If we set our machine that computes f to emit $f(1)$, $f(2)$, and so on (or even run it in parallel with itself if we are not assuming that f is total), then if n is indeed a value of f, we will learn this sooner or later; but if it isn't, this process will never tell us. However, this does at least give us a *verification procedure*: we can detect membership of f"$\mathbb{N}$ in these circumstances even though we are not promised an exclusion procedure. Thus the natural idea seems to be that of a *semi*decidable set: one for which membership can be confirmed in finite time.

But is this the only way we can exploit computable functions to get a concept of semidecidable set? Being the range of a computable function seems a pretty good explication of the concept of a semideciable set, but then being the set of arguments on which a computable function halts – $\{n : f(n) \downarrow\}$ seems pretty good too. After all, if $f(n) \downarrow$, then we will certainly learn this in finite time. Fortunately for us, all obvious attempts to capture the concept of semidecidable set using these ideas give the same result.

REMARK **43** *The following conditions on a set* $X \subseteq \mathbb{N}$ *are equivalent.*
   (i) X is the range of a μ-recursive function.
   (ii) X is the set of naturals on which a μ-recursive function is defined.
   (iii) X is the range of a μ-recursive function that happens to be total.

Proof: (i) $\rightarrow$ (iii). The converse is obvious since (iii) is a special case of (i). The key idea here is that of finite but unbounded parallelism, an important idea that we will now explain.

Suppose X is the range of a computable function f and $\mathcal{M}$ is a machine that computes f. The idea of autoparallelism is that at stage n we run $\mathcal{M}$ with input `fst` n for `snd` n steps. When we do this with a machine, the effect is that we keep trying the machine with all inputs, continually breaking off and revisiting old inputs – and continually starting computations on new, later inputs – so that every computation is given infinitely many chances to halt. Of course once a computation with input k has halted, we do not revisit it. Therefore, at stage n, if **fst** n is an input that has already halted, we procede at once to stage $n + 1$.

This autoparallelism is really a breadth-first search through all the computations that $\mathcal{M}$ is capable of.

We run $\mathcal{M}$ in parallel with itself as just described and declare g to be the function that sends an input n to the nth thing output by M when run in parallel with itself. g is total, and clearly it outputs all and only the members of X. (I am ignoring the case where X is finite: it is a miniexercise to check for yourselves that this is safe!)

(i) $\to$ (ii) Given a machine $\mathcal{M}$ that outputs members of X, we can build a machine $\mathcal{M}'$ that on being given a number n runs $\mathcal{M}$ in parallel with itself as above until it produces the output n: $\mathcal{M}'$ then outputs 0, say (it does not matter). $\mathcal{M}'$ is then a machine that halts on members of X and nothing else.

(ii) $\to$ (i) Given a machine $\mathcal{M}$ that halts on members of X, we can build a machine that outputs members of X by simply trapping the output of $\mathcal{M}$ and outputting the input instead of the output. ∎

EXERCISE 82 *Show that one can take the total computable function that emits members of X in part (iii) of remark 43 to be one-to-one.*

Incurable optimists might hope that this autoparallelism might give us a cure to the problem discussed on page 131 in section 6.2. After all, there is always the possibility of running g in parallel with itself. Will this help? Although that will turn up an input y to g s.t. $g(y, \vec{x}) = n$ if there is one, there is no reason to suppose it will turn up the smallest. Indeed, quite which one it turns up will depend on how we have implemented the autoparallel algorithm, so even which functions turn out to be computable would depend on how we implement the algorithm! This is clearly intolerable.

EXERCISE 83 *Suppose that ϕ is a partial function of two arguments.*

(i) *Show that there is a partial computable function ψ of one argument such that, for each m, if there are x with $\phi(x, m) = 0$, then $\phi(\psi(m), m) = 0$. If there are no such x, is your $\psi(m)$ defined?*

(ii) *Show that it is not always possible to take $\psi(m) = \mu x.(\phi(x, m) = 0)$.*

See also exercise 90 in section 6.6.

DEFINITION 44 *A set satisfying the conditions in remark 43 is* **semi-decidable**[1] *A set X is* **decidable** *if X and $\mathbb{N} \setminus X$ are both semidecidable.*

Just as "computable function" is better than "recursive function" because recursion is not always prominent in the declaration of a computable function, so "decidable set" is better than "recursive set" (the old terminology), since "recursive set" would suggest that there also ought to be "primitive recursive set" – you are one if you are the range of a primitive recursive function. But in fact

EXERCISE 84

  (i) *Every decidable set is the range of a primitive recursive function. (Hint: Use autoparallelism and Kleene's T-function.)*

  (ii) *Show that condition (i) of remark 43 can be strengthened to "X is f "Y for some computable f and decidable Y".*

Note the parallel between the idea of a *regular language*, which is the set of strings accepted by a finite-state machine, and the idea of a *semidecidable set*, which is the set of natural numbers on which a Turing machine will halt.

If X is semidecidable, it is f "$\mathbb{N}$ for some total computable f, so whenever $n \in X$ there is $k \in \mathbb{N}$ and a finite computation verifying that $f(k) = n$, so that $n \in X$. This finite computation should be thought of as a *proof* or *certificate* in the sense of section 2.1.7, so a semidecidable set of naturals can be thought of as a subset of $\mathbb{N}$ that happens to be a rectype in its own right. Indeed, we can take this further: by means of gnumbering every finitely presented rectype can be thought of as a semidecidable set.

We are now in a position to give a slightly more natural version of definition 18. An axiomatisable theory is one with a set of axioms whose gnumbers form a semidecidable set. (It is assumed that the theory only has finitely many rules of inference. Without that condition, any theory at all could be axiomatisable as follows: take an empty set of axioms,

---

[1] The old terminology is 'recursively enumerable', which is gradually going out of fashion. That notation arises because any set of natural numbers can be enumerated (and *enumerable* or *denumerable* are old words for 'countable'), but not necessarily by a computable function. If the set is enumerated by a *recursive* function, it is *recursively* enumerable. Bear in mind too that in some of the literature 'semidecidable' is used to mean 'semidecidable and not decidable'.

and for each theorem have a nullary rule of inference whose conclusion is that theorem.)

EXERCISE 85

(i) *Show that any theory that can be axiomatised with a set of axioms that is semidecidable also has a set of axioms that form a decidable set. (Beware of this question: its proof is very silly.)*

(ii) *Since propositional logic is decidable, the set of falsifiable propositional formulæ over an alphabet is also semidecidable, so it is a rectype. Give a presentation.*

It might be felt that the following definition of decidable sets is more natural: X is decidable iff there is a total computable function $f : \mathbb{N} \to \{0,1\}$ such that $X = f^{-1}\text{``}\{1\}$.

EXERCISE 86 *Check that a set is decidable iff there is a total computable function $f : \mathbb{N} \to \{0,1\}$ such that $X = f^{-1}\text{``}\{1\}$.*

The original definition looks more cumbersome and long-winded, but if one starts with the definition of decidable sets given in exercise 86, it is much harder to motivate the concept of semidecidable set and the connection between the two ideas is less clear.

I emphasised that concentrating on partial functions was the conceptual breakthrough: it was that that enabled us to prove the completeness theorem for computable partial functions. Quite how big a mess we would have got into if we had stuck with total functions is shown by the diagonal argument:

THEOREM 45 *The set of gnumbers of total computable functions is not semidecidable.*

Proof: Suppose the set of gnumbers of machines that compute total functions were semidecidable. Then there would be a total computable function f whose values are precisely the gnumbers of machines that compute total functions. Indeed, let f_n be the function computed by the machine whose gnumber is $f(n)$. Now consider the function $\lambda n.f_n(n)+1$. This function is total computable and should therefore be f_m for some m. But it cannot be f_m, because its value for argument m is $f_m(m) + 1$ and not $f_m(m)$. ∎

This should not come as a surprise. Ask yourself: if I am given the gnumber of a machine, can I confirm in finite time that the function computed by that machine is total? At the very least, it is obvious that there is no *straightforward* way of confirming this in finite time. So one should not be surprised that there is in fact no way at all of doing it – in finite time.

From now on we say "computable" instead of "μ-recursive". You may also hear people saying "general recursive" or "partial recursive", which mean the same thing. Confusingly, you will also hear people talk about functions being *partial recursive* in contrast to being *total recursive*. A set is **decidable** if its **characteristic function**[1] is computable.

DEFINITION 46 *The characteristic function χ_A of $A \subseteq \mathbb{N}$ is*

$$\lambda n.\ \text{if}\ x \in A\ \text{then}\ 1\ \text{else}\ 0.$$

(The Greek letter 'χ' is the first letter of the Greek word for 'character'.)

6.4 The undecidablity of the halting problem

The set of register machine programs is countable because of the prime powers trick. The set of all subsets of $\mathbb{N}$ is not, because of Cantor's theorem (theorem 6). There simply are not enough register machine programs to go round: inevitably some subsets of $\mathbb{N}$ are going to be undecidable. In fact, *almost all* of them are, in the sense that there is the same number of subsets of $\mathbb{N}$ as there are undecidable subsets. This argument is nonconstructive and does not actually exhibit a subset of $\mathbb{N}$ that is not decidable, but we can do that too.

Suppose we had a machine that, on being given a natural number n, decoded it (using the primitive recursive unpairing functions alluded to on page 128) into fst n and snd n (n_1 and n_2 for short), and then $\downarrow= 0$ if the n_1th machine halts when given input n_2 and $\downarrow= 1$ otherwise.

We can tweak this machine (by using something to trap the output) to get something with the following behaviour: on being given n, it decodes it into n_1 and n_2 (fst and snd of n) and then $\downarrow= 1$ if the n_1th machine diverges on input n_2 (just as before) but diverges if the n_1th machine halts when given input n_2.

Front-end onto *this* machine a machine that accepts an input x and outputs $\text{pair}(x, x)$. We now have a machine with the following behaviour.

---

[1] In other traditions they are sometimes called **indicator functions**.

On being given n, it tests to see whether or not the nth machine halts with input n. If it does, it goes into an infinite loop (diverges); if not, it halts with output 1.

This machine is the n_0th, say. What happens if we give it n_0 as input? Does it halt? Well, it halts iff the n_0th machine loops when given input n_0. But it *is* the n_0th machine itself!

Formally we can write $\phi_{n_0}(n_0) \downarrow$ iff (by definition of ϕ_{n_0}) $\phi_{n_0}(n_0) \uparrow$. Notice the similarity with the proof of Cantor's theorem (section 2.1.6.2).

What assumption can we discard to escape from this contradiction? Clearly we cannot discard the two steps that involve just trapping output and front-ending something innocent onto the hypothesised initial machine. The culprit can only be that hypothesised machine itself! So we have proved

THEOREM 47 *The set of numbers* $\mathtt{pair}(p, d)$ *such that p halts on d is not decidable.*

Though it is obviously semidecidable!

6.4.1 Rice's theorem

Theorems 45 and 47 are manifestations of a general phenomenon, captured by Rice's theorem. (Though theorem 45 is actually slightly stronger than a special case of Rice's theorem.) We prove a number of results *en route* to Rice's theorem.

THEOREM 48 *(The S-m-n theorem) There is a computable total function S such that*

$$\phi_e(a, b) = \phi_{S(e,b)}(a),$$

and so on for higher degrees (more parameters).

This is a corollary of the equality between μ-recursiveness and computability by register machines: one can easily tweak a machine for computing $\lambda ab.\phi_e(a, b)$ into a machine that, on being given a, outputs a description of a machine to compute $\lambda b.\phi_e(a, b)$.

In turn we get a corollary:

COROLLARY 49 *(The fixed point theorem) Let $h : \mathbb{N} \to \mathbb{N}$ be a total computable function. Then there is n such that $\phi_n = \phi_{h(n)}$.*

Proof: Consider the map

$$\texttt{pair}(e, x) \mapsto \phi_{h(S(e,e))}(x).$$

This is computable and is therefore computed by the ath machine, for some a. Set $n = S(a, a)$. Then

$$\phi_n(x) =^1 \phi_{S(a,a)}(x) =^2 \phi_a(a, x) =^3 \phi_{h(S(a,a))}(x) =^4 \phi_{h(n)}(x)$$

Equation (1) holds because $n = S(a, a)$; (2) holds by definition of S; (3) holds by definition of a and (4) holds by definition of n.

Notice that we need h to be total computable.

There is a powerful corollary of this that is a sort of omnibus undecidablity theorem.

THEOREM 50 *(Rice's theorem) Let A be a nonempty proper subset of the set of all recursive functions of one variable. Then $\{n : \phi_n \in A\}$ is not decidable.*

Proof: Suppose $\chi(A)$ is computable; we will deduce a contradiction.

Find naturals a and b so that $\phi_a \in A$ and $\phi_b \notin A$. (Not only are there such a and b, but we can find them because $\chi(A)$ is computable.) Since $\chi(A)$ is computable the following function is also computable:

$$g(n) := \quad \text{if} \ \ \phi_n \in A \ \text{then} \ b \ \text{else} \ a$$

("wrong way round"!). By corollary 49 there must now be a number n such that $\phi_n = \phi_{g(n)}$. Is ϕ_n in A?

If it is, then (i) $\phi_{g(n)} \in A$ (since $\phi_n = \phi_{g(n)}$) and (ii) $g(n) = b$ by construction of g. But if $\phi_n = \phi_{g(n)}$, then $\phi_{g(n)} \in A$, whence $g(g(n)) = b$ (by construction of g). Now $g(n) = b$, so $g(g(n)) = g(b) = a$. This contradiction shows that $\phi_n \notin A$.

Now try $\phi_n \notin A$. We have (i) $\phi_{g(n)} \notin A$ (since $\phi_n = \phi_{g(n)}$) and (ii) $g(n) = a$ by construction of g. But if $\phi_n = \phi_{g(n)}$, then $\phi_{g(n)} \notin A$, whence $g(g(n)) = a$ (by construction of g). Now $g(n) = a$, so $g(g(n)) = g(a) = b$. This gives a contradiction too, so we must drop our assumption that A was decidable. ∎

This theorem is very deep and very important, but the moral it brings is very easy to grasp. It tells us that we can never find algorithms to answer questions about the *behaviour* of programs ("Does it halt on this input?"; "Does it always emit even numbers when it does halt?") on the basis of information purely about the *syntax* of programs ("Every variable occurs an even number of times"). In general, if you want to know anything about the behaviour of a program, you may be lucky and succeed in the short term and in a small number of cases, but in the long run you cannot do better than by just running it.

In particular it has the consequence that it is not decidable whether or not two programs compute the same function(-in-extension). This makes it particularly important to bear in mind that the theory of computable functions is in the first instance a study of function declarations (functions-in-intension) rather than function graphs.

6.5 Relative computability

Quite early on in the development of the theory of computable functions people noticed that the techniques developed to study computability generalise naturally to enable one to study *relative computability*, what is termed so evocatively *computation relative to an oracle*. All one has to do is enhance the machine architecture by adding a state in which the machine consults an oracle, which will be a subset of $\mathbb{N}$ or a function $\mathbb{N} \to \mathbb{N}$. This leads one naturally to the study of equivalence classes of functions $\mathbb{N} \rightharpoonup \mathbb{N}$ under the relation of being-equally-computable.

6.6 Exercises

EXERCISE 87 *Define the terms* primitive recursive function, partial recursive function, *and* total computable function.

Ackermann's function is defined as follows:

$$A(0,y) := y+1; \; A(x+1,0) := A(x,1); \; A(x+1,y+1) := A(x,A(x+1,y)).$$

For each n define $f_n(y) := A(n,y)$. Show that for all $n \geq 0$, $f_{n+1}(y) = f_n^{y+1}(1)$, and deduce that each f_n is primitive recursive. Why does this mean that the Ackermann function is total computable?

EXERCISE 88 *For which of the following functions-in-intension are there computable functions-in-intension with the same extension?*

    (i) $\lambda x.$ *if there is somewhere in the decimal expansion of π a string of exactly x 7's, then 0, else 1;*

    (ii) $\lambda x.$ *if there is somewhere in the decimal expansion of π a string of at least x 7's, then 0, else 1;*

    (iii) $\lambda k.$ *the least n such that all but finitely many natural numbers are the sum of at most n kth powers.*

EXERCISE 89 *Recall from page 7 the idea of the graph of a function. Show that the graph of a total computable function $f : \mathbb{N}^n \to \mathbb{N}$ is a decidable subset of $\mathbb{N}^{n+1}$. Is the graph of a partial computable function decidable?*

EXERCISE 90 *Suppose that f is a total computable function satisfying $\forall n.f(n) \leq f(n+1)$. Show that the range of f is a decidable set. [Hint: The range of f is either finite or infinite; consider these two cases separately. Be warned that your proof will not be constructive!]*

EXERCISE 91 *A **box of tiles** is a set of rectangular tiles, all of the same size. The tiles have an orientation (top and bottom, left and right) and the edges have colours. The idea is to use the tiles in the box to tile the plane, subject to rules about which colours can be placed adjacent to which, and each box comes with such a set of rules. (Naturally every set of rules includes all the obvious things, like, a bottom edge can only go next to a top edge, and so on.) So of course the box has infinitely many tiles in it. Nevertheless, the tiles can only be of finitely many kinds. (It is a bit like a scrabble set: only 27 letters but many tokens of each)*

With some boxes one can tile the plane; with some one cannot. Sketch how to gnumber boxes and explain why the set of gnumbers of boxes that cannot tile the plane is a semidecidable set.

EXERCISE 92 *Let $f : \mathbb{N} \to \mathbb{N}$ be a strictly order-preserving total computable function. Construct a semidecidable subset A of $\mathbb{N}$ such that (i) for all e, if the domain $\mathrm{Dom}(\phi_e)$ of the eth partial computable function is infinite, then $\mathrm{Dom}(\phi_e) \cap A \neq \emptyset$ and (ii) there are at most e elements less than $f(e)$.*

Deduce that there is a semidecidable set B such that $\mathbb{N} \setminus B$ is infinite and contains no infinite semidecidable subset.

EXERCISE 93 *Is it possible to decide, given that ϕ_e is total, whether or not*

   (i) $\forall n.\phi_e(n) = 0$?
   (ii) $\exists n.\phi_e(n) \leq \phi_e(n+1)$?
   (iii) $\exists n.\phi_e(n) \geq \phi_e(n+1)$?

EXERCISE 94 *Any natural substitution function S will have $S(e,n) > e$ and $S(e,n) > n$ for all e and n. Deduce that, for any (partial) computable f, there are infinitely many e with $\phi_e = f$.*

EXERCISE 95 *Show that the following sets are not decidable.*

(i) $\{e : \phi_e$ *everywhere undefined* $\}$.   *(ii)* $\{e : \phi_e$ *is total* $\}$.

(iii) $\{e : \forall i < e.\,(\phi_e(i) \downarrow)\}$.   *(iv)* $\{e : \forall i.\,(\phi_e(i) \downarrow \rightarrow i < e)\}$.

EXERCISE 96 *Is the following true or false? If* $h : \mathbb{N} \rightarrow \mathbb{N}$ *is total computable, then there is an* e *such that* ϕ_e *is total and* $\phi_e = \phi_{h(e)}$.

EXERCISE 97 *Suppose that* $f, g : \mathbb{N}^2 \rightarrow \mathbb{N}$ *are total computable. Show that there exist* i, j *with* $\phi_i = \phi_{f(i,j)}$ *and* $\phi_j = \phi_{g(i,j)}$. *[Hint: Show first that there is a total computable* h *with* $\phi_{h(i)} = \phi_{g(i,h(i))}$.*] (Hard)*

EXERCISE 98 *Show that* $A \subseteq \mathbb{N}$ *is semidecidable just when it is of the form* $\{n \in \mathbb{N} : \exists m.R(n,m)\}$ *for some recursive predicate* R.

EXERCISE 99 *Which of the sets* R *and their complements* $\mathbb{N} \setminus R$ *from exercise 95 are semidecidable?*

EXERCISE 100 *By considering enumerations of the partial computable functions, find a partial computable function that cannot be extended to a total computable function.*

EXERCISE 101 *Show that for any corruptible operating system there can be no program* IS-SAFE *that, when given program p and data d, says "yes" if p applied to d does not corrupt the operating system and "no", otherwise.*

EXERCISE 102 *(For lambda hackers only) The Church numeral* **n** *is that lambda term representing the function that – when given a function* f *– returns the function that does* f *to its argument n times. Thus Church numeral 1 is the identity. Church numeral 0 is K of the identity. Find a lambda term for successor. How do we implement multiplication and addition?*

EXERCISE 103 *(For lambda hackers only) Using the pairing and unpairing lambda terms you discovered earlier and your answer to question 78(ii), show that any primitive recursive function can be represented by a lambda term acting on Church numerals.*

EXERCISE 104 *What might a decidable partition of* $\mathbb{N}$ *be? Exhibit a decidable partition of the set of unordered triples from* $\mathbb{N}$ *such that any set monochromatic for it can be used to solve the halting problem.*

7

Ordinals

The word 'ordinal' has been used for years to denote a kind of number word: there are ordinals and cardinals. Cardinals are words like 'one', 'two', 'three'; ordinals are words like 'first', 'second', 'third'. Although some of the original nature of the difference has been lost in the process of having these words appropriated by mathematics, a significant and important part remains: ordinal numbers allude to order, and to positions in a sequence. Happily, the best introduction to these ideas is by way of their historically first application.

For reasons we cannot go into here, Cantor was interested in the complexity of closed sets in $\Re$. A closed set might be a **perfect** closed set (a union of closed intervals, so that every point is a limit point), or it might have some isolated points. If one removes the isolated points from a closed set, one might get a perfect set, but one might not. It might be that once one removes all the isolated points from a closed set, a point that had not been isolated before now becomes isolated. One measures the complexity of a closed set by the number of times one has to perform this operation of deleting isolated points to obtain a perfect closed set. The interesting feature is that, even if one performs this deletion infinitely often, one is not assured of obtaining a perfect closed set. It is not difficult to construct a closed set containing a point x that is the limit of a sequence $\{x_n\}$ where x_k becomes isolated at stage k. x itself then never gets deleted, but it becomes isolated after infinitely many stages. But all is not lost. After we have performed the deletion operation infinitely many times, we can look at what is left and perform the deletion operation on that, and thereby continue the process transfinitely. One can hope that eventually a perfect closed set is reached.

Let us now stand back and ask ourselves what it was about this scenario that made it possible to apply this operation transfinitely. All that

147

was needed was that there should be a monotone (increasing or decreasing, it makes no difference) function from some poset into itself, which is continuous, *so that we have a well-defined notion of what happens at a limit stage.*

Ordinals are now invoked as that kind of number that counts stages. This is turn naturally generates them as a rectype: for any stage there is a just-next stage, and for any increasing sequence of stages there is a supremum stage. The class of stages thus forms a rectype whose engendering relation is a well-order (see theorem 4).

Since the set of stages of any construction indexed in this way is naturally well-ordered by the engendering relation of the rectype of stages of the construction, one is led to consider the isomorphism types of well-orderings. This is another way of thinking of ordinals. These two ways are complementary, and both are right. We should now cast our minds back to the two ways we have of thinking of natural numbers. We can think of them as sizes of finite sets, or we can think of them as the members of a certain inductively defined set. These two ways of thinking about natural numbers correspond to the two ways of thinking of ordinals. Ordinals can be thought of as isomorphism classes of well-orderings, or they can be thought of as members of a rectype. We will discuss the two approaches in turn.

7.1 Ordinals as a rectype

Lowercase Greek letters are used to range over ordinals. Its use in λ-calculus notwithstanding, the letter 'λ' is always liable to a variable ranging over *limit* ordinals in the way that in A-level analysis 'x' and 'y' are ordinate and abcissa or input and output variables.

We are going to derive ordinal arithmetic in a fairly relaxed and informal way from ordinals constructed as a rectype in a way suggested by the following ML-style pseudocode:

```
new_data_type ordinal = 0
                      | succ of ordinal
                      | sup of (chain-of ordinal)
```

The occurrence of the word 'chain' in the second clause of course presupposes an ordering, so we must come clean on that, by defining $\leq_{On}$ recursively as follows:

DEFINITION **51** *We start by noting that (as with* $\mathbb{N}$*)* succ *is understood to have no fixed points.*

$\alpha \leq_{On} \beta \to \beta \leq_{On} \gamma \to \alpha \leq_{On} \gamma.$

$\alpha \leq_{On} \alpha.$

$\alpha \leq_{On} \beta \leq_{On} \alpha \to \alpha = \beta.$

$0 \leq_{On} \alpha.$

$\alpha \leq_{On}$ succ $\alpha.$

$\alpha \leq_{On} \beta \to$ succ $\alpha \leq_{On}$ succ $\beta.$

$\alpha \in X \to \alpha \leq_{On}$ sup $X.$

$(\forall \alpha \in X)(\alpha \leq \beta) \to$ sup $X \leq_{On} \beta.$

Naturally we will also want '$\alpha <_{On} \beta$' *as short for* '$\alpha \leq_{On} \beta \wedge \beta \not\leq_{On} \alpha$'.

We will now write '*On*' for the class of all ordinals. The ordinals are very nearly a complete poset, but not quite. The presence of the succ operator prevents there being a top element, but every *bounded* chain has a least upper bound. The following exercise is really only for enthusiasts: it's a bit fiddly.

EXERCISE **105** *Show that* $<_{On}$ *is a well-ordering. Hint: Recycle the proof of Witt's theorem (theorem 10) to show it is a total ordering, then use theorem 4.*

Declaring the ordinals like this is a kind of Indian rope trick, but it does at least give us a picture of ordinals as things that count stages.

7.1.1 Operations on ordinals

Now we can give some recursive definitions of the obvious operations, starting with addition.

DEFINITION **52**

$\alpha + 0 := \alpha;$

$\alpha +$ succ $\beta :=$ succ $(\alpha + \beta);$

$\alpha +$ sup $X :=$ sup $\{\alpha + \beta : \beta \in X\}.$

EXERCISE **106** *For all ordinals* α *and* β, $\alpha \leq_{On} \beta$ *iff* $(\exists \gamma)(\alpha + \gamma = \beta).$

Now we can procede to define multiplication:

DEFINITION 53

$\alpha \times 0 := 0;$

$\alpha \times \text{succ } \beta := (\alpha \times \beta) + \alpha;$

$\alpha \times \sup X := \sup \{\alpha \times \beta : \beta \in X\};$

and exponentiation:

DEFINITION 54

$\alpha^0 := \text{succ } 0;$

$\alpha^{(\text{succ } \beta)} := (\alpha^\beta) \times \alpha;$

$\alpha^{(\sup X)} := \sup \{\alpha^\beta : \beta \in X\}.$

Given these definitions, it is clear that addition on the right, multiplication on the right and exponentiation on the right, namely, the
functions $\lambda\alpha.(\beta + \alpha)$, $\lambda\alpha.(\beta \times \alpha)$ and $\lambda\alpha.(\beta^\alpha)$ are – for each ordinal
β – *continuous* in the sense in which the ordinals are (very nearly) a
chain-complete poset.

EXERCISE 107 *Look again at part* (v) *of exercise 4 3.3.2.1.*
, which shows that these operations are noncommutative.

(i) *Give examples to show that addition and multiplication on the
left are not commutative.*

(ii) *Give an example to show that* $\lambda\alpha.\alpha^2$ *is not continuous.*

(iii) *Which of the following are true for all* α, β *and* γ?

(a) $(\alpha \times \beta)^\gamma = \alpha^\gamma \times \beta^\gamma;$

(b) $\gamma^{(\alpha+\beta)} = \gamma^\alpha \times \gamma^\beta;$

(c) $(\alpha + \beta) \times \gamma = \alpha \times \gamma + \beta \times \gamma;$

(d) $\gamma \times (\alpha + \beta) = \gamma \times \alpha + \gamma \times \beta.$

*Prove the true assertions and give counterexamples to the false
assertions.*

We will need later a notion of ordinal **subtraction**. $\alpha - \beta$ is the length
of a well-ordering obtained from a well-ordering of length α by chopping
off an initial segment of length β.

EXERCISE 108 *Give a recursive definition of ordinal subtraction, and
prove that your definition obeys* $\beta + (\alpha - \beta) = \alpha$.

We have already invoked a concept of continuity of functions from On
(or $(On \times On)$) to On. For the following definition we need to hark back

to the idea of the order topology: a set of ordinals is closed iff it contains all its limit points.

DEFINITION 55 *A* **clubset** *is a* **CL***osed and* **U***n***B***ounded set, or, alternatively, the range of a total continuous function. (Sometimes it is called a* **normal** *function.)*

Thus a normal function is strictly increasing and continuous. It is obvious that every normal function has a fixed point. If f is normal, then $\sup\{f^n \alpha : n \in \mathbb{N}\}$ is the least fixed point for f above α. In fact:

LEMMA 56 *The function enumerating the set of fixed points of a normal function is also normal.*

Proof: See your answer to exercise 3.1.3 22. ∎

7.1.2 Cantor's normal form theorem

To prove Cantor's normal form theorem we will need to make frequent use of the following important triviality.

REMARK 57 *If $f : On \to On$ is normal, then for every $\alpha \in On$ there is a maximal $\beta \in On$ such that $f(\beta) \le \alpha$.*

Proof: Consider the set of β such that $f(\beta) \le \alpha$, and let β_0 be its sup. By continuity of f, $f(\beta_0) \le \alpha$ and is clearly maximal with this property. ∎

This enables us to prove a normal form theorem for ordinal notations.

If $\alpha < \beta$, then there is a largest γ such that $\alpha^\gamma \le \beta$ by remark 57. Call this ordinal γ_0. Then $\alpha^{\gamma_0} \le \beta$. If $\alpha^{\gamma_0} = \beta$, we stop there.

Now consider the case where $\alpha^{\gamma_0} < \beta$. By remark 57, there is a maximal θ such that $\alpha^{\gamma_0} \cdot \theta \le \beta$. Call it θ_0. If $\alpha^{\gamma_0} \cdot \theta_0 = \beta$, we stop there, so suppose $\alpha^{\gamma_0} \cdot \theta_0 < \beta$. Now $\beta = \alpha^{\gamma_0} \cdot \theta_0 + \delta_0$ for some δ_0 (remember dfn of $<_{On}$).

What we have proved is that, given ordinals $\alpha < \beta$, we can express β as $\alpha^{\gamma_0} \cdot \theta_0 + \delta_0$ with γ_0 and θ_0 maximal. If $\delta_0 < \alpha$, we stop. However, if $\delta_0 > \alpha$, we continue, by repeating the above process with α and δ_0.

What happens if we do this? We then have $\delta = \alpha^{\gamma_1} \cdot \theta_1 + \delta_1$, which is to say

$$\beta = \alpha^{\gamma_0} \cdot \theta_0 + \alpha^{\gamma_1} \cdot \theta_1 + \delta_1.$$

One thing we can be sure of is that $\gamma_0 > \gamma_1$. This follows from the maximality of θ_0. Therefore, when we repeat the process to obtain:

$$\beta = \alpha^{\gamma_0} \cdot \theta_0 + \alpha^{\gamma_1} \cdot \theta_1 + \alpha^{\gamma_2} \cdot \theta_2 + \cdots \alpha^{\gamma_n} \cdot \theta_n + \cdots$$

we know that the expression can only be finitely long, because the sequence of ordinals $\{\gamma_0 > \gamma_1 > \gamma_2 > \gamma_n \ldots\}$ is a descending sequence of ordinals and must be finite, because $<_{On}$ is well-founded.

So we have proved this:

THEOREM 58 *For all β and all $\alpha < \beta$, there are $\gamma_0 > \cdots > \gamma_n$ and $\theta_0 \cdots \theta_n$ such that*

$$\beta = \alpha^{\gamma_0} \cdot \theta_0 + \alpha^{\gamma_1} \cdot \theta_1 + \alpha^{\gamma_2} \cdot \theta_2 + \cdots \alpha^{\gamma_n} \cdot \theta_n + \cdots . \blacksquare$$

If $\alpha = \omega$, all the θ_n are finite. (If any of them were bigger than ω, then the corresponding γ_n would not have been maximal.) This means that we can actually take each θ_n to be 1, by allowing finitely many repeats.

Quite how useful this fact is when dealing with an arbitrary ordinal β will depend on β. After all, if $\beta = \omega^\beta$, then all that Cantor's normal form theorem will tell us if we run the algorithm with ω and β is that this is, indeed, the case. Ordinals β such that $\beta = \omega^\beta$ are around in plenty. They are called ϵ-**numbers**. They are moderately important because, if β is an ϵ-number, then the ordinals below β are closed under exponentiation. The smallest ϵ-number is called 'ϵ_0'. For the moment, what concerns us about ϵ_0 is that, if we look at the proof of Cantor's normal form theorem in the case where β is an ordinal below ϵ_0 and $\alpha = \omega$, the result is something sensible. This is because, ϵ_0 being the *least* fixed point of $\lambda\alpha.\omega^\alpha$, if we apply the technique of remark 57 to some $\alpha < \epsilon_0$, the output of this process must be an expression containing ordinals below α.

7.2 Ordinals from well-orderings

A well-ordering is a well-founded strict total order. Ordinals will emerge as the kind of number that measures the length of well-orderings.

The result of adding an extra element to the end of a well-ordering is of course another well-ordering: this corresponds to the successor constructor of the rectype of ordinals. The result of concatenating one well-ordering on the end of another is a well-ordering, and this will eventually correspond to ordinal addition. The lexicographic product of two well-orderings is a well-ordering, as we saw in exercise 5, and lexicographic product will eventually correspond to ordinal multiplication.

However, we also want an operation on well-orderings that corresponds to the **sup** constructor on ordinals, and care is needed. It is not true that the direct limit of an increasing sequence of well-orderings is a well-ordering: let $\mathcal{A}_n$ be $\langle \{x \in Z : -n < x < 0\}, < \rangle$ with the inclusion mapping between $\mathcal{A}_n$ and $\mathcal{A}_{n+1}$. The direct limit is the negative integers ordered by magnitude, and this is not a well-order. For a direct limit of well-orderings to be a well-ordering we need the "new" elements to be put "on the end" in a sense that we will now make precise.

DEFINITION 59 *Given two binary structures $\langle A, R \rangle$ and $\langle B, S \rangle$, we say $\langle B, S \rangle$ is an **end-extension** of $\langle A, R \rangle$ if $A \subseteq B$ and $R \subseteq S$, and whenever $y \in A$ and xSy then $x \in A$ too.*

(You will already have discovered this concept in your answer to exercise (ii) on page 61, and in your answer to exercise 19 on page 56.)

For the moment we will be primarily interested in the case where $\langle A, R \rangle$ and $\langle B, S \rangle$ are partial orders – indeed well-orders, and in those circumstances the picture is easy to paint in slogan form: the new members all come after the old members. We will later also be interested in the case where both R and S are $\in$, and in this case the slogan is "new sets – yes; new members of old sets – no!".

In general, the concept of end-extension is not very useful, except in connection with models of a theory of a rectype, and then only when the end-extension is with respect to the engendering relation of the rectype.

We can prove

PROPOSITION 60 *The direct limit of a chain of well-orderings under end-extension is a well-ordering.*

Proof: Let $\mathcal{A}$ be the union of a chain $\{\mathcal{A}_i : i \in I\}$ of well-orderings where for $i < j$, $\mathcal{A}_j$ is an end-extension of $\mathcal{A}_i$. Let A' be a subset of A, the carrier set of $\mathcal{A}$. We want A' to have a least element. For sufficiently large $i \in I$, A' must meet A_i, the carrier set of $\mathcal{A}_i$, and so $A' \cap A_i$ must have a least element, a_i, say. It remains to be shown that a_i does not depend on i. Suppose $i < j$ and $a_i \neq a_j$. How can this be? $A_j \cap A'$ is a superset of $A_i \cap A'$ and so must have an inf no bigger than that of $A_i \cap A'$. This tells us that $a_j \leq a_i$. We cannot have $a_j < a_i$ because everything in $A_j \setminus A_i$ comes *after* everything in A_i. ∎

This last result sounds as if it is in need of strengthening: we want all directed sups of well-orderings (under end-extension) to be well-orderings, not just unions of chains.

DEFINITION 61 *If $c_1 \subseteq c_2$ are two subsets of a poset with $(\forall p \in c_2)(\exists p' \in c_1)(p \leq p')$, we say c_1 is **cofinal** in c_2.*

Typically, when this definition is invoked, c_2 is a chain, as in the standard result that (assuming the axiom of choice) every directed poset has a cofinal chain (which was exercise (xi) on page 61), so this application will give us the strengthening of 60 that we seek. But in fact we can get it even without AC, since all well-orderings are comparable in point of length, as we shall now see.

PROPOSITION 62 *Given two well-orderings $\mathcal{A} = \langle A, \leq_A \rangle$ and $\mathcal{B} = \langle B, \leq_B \rangle$, there is a unique isomorphism between one and an initial segment of the other.*

Proof: We define the isomorphism by recursion. The idea is that we pair off the $\leq_A$-first member of A with the $\leq_B$-first member of B, and thereafter we pair the $\leq_A$-first thing in A (that has not already been used) with the $\leq_B$-first thing in B (that has not already been used). There always *is* a first thing that has not already been used since $\mathcal{A}$ and $\mathcal{B}$ are well-orderings, so every subset has a first member.

We consider the class of *partial isomorphisms* between $\mathcal{A}$ and $\mathcal{B}$. These are what you might expect: isomorphisms between an initial segment of $\mathcal{A}$ and an initial segment of $\mathcal{B}$. We will also need the concept of *two partial isomorphisms agreeing on their intersection*. We then prove by induction on $\mathcal{A}$ that, for all $a \in A$, if i and j are partial isomorphisms $A \to B$ that are defined at a (i.e., $i(a)$ and $j(a)$ are both defined) then i and j agree on a, that is, $i(a) = j(a)$. For suppose not. Let a be the $\leq_A$-least element of A such that there are partial isomorphisms i, j from $A \to B$ s.t. $i(a) \neq j(a)$. This must mean that $\mathcal{B}$ has two elements $i(a)$ and $j(a)$ that are equally plausible as mates for a. But this cannot be, since $\mathcal{B}$ is a well-ordering and so one of $i(a)$, and $j(a)$ must come earlier than the other and be the only fit mate for a. This means that we can sensibly introduce a notation a_B for the element of $\mathcal{B}$ that a must be paired with. We can do the same for $\mathcal{B}$, so that to $b \in B$ there should correspond a b_A. Notice that there is no reason to suppose that an arbitrary element of $\mathcal{A}$ is in the range of any partial isomorphism: $\mathcal{A}$ might be much longer than $\mathcal{B}$, or vice versa. This concentrates our minds on the two functions $\lambda a \in \mathcal{A}.a_{\mathcal{B}}$ and $\lambda b \in \mathcal{B}.b_{\mathcal{A}}$. One or the other might be partial instead of total (if $\mathcal{A}$ is longer than $\mathcal{B}$, then $\lambda a \in \mathcal{A}.a_{\mathcal{B}}$ will be partial), but they cannot both be partial. (If they are, they can both be extended.) ∎

COROLLARY 63 *No well-ordering is the same length as any of its proper initial segments.*

Proof: Apply proposition 62 to the situation where $\mathcal{A}$ and $\mathcal{B}$ are the same well-ordering. It tells us that there is a unique isomorphism between $\mathcal{A}$ and some initial segment of $\mathcal{A}$. If there is only one isomophism, there is only one initial segment, and since $\mathcal{A} \simeq \mathcal{A}$, that initial segment must be $\mathcal{A}$ itself. ∎

We can define the rectype of well-orderings as generated from the empty poset by the operations of adding new elements on the end and of taking sups of chains (where the order relation is end-extension). Of course it is more usual (and, mostly, more useful) to say that a relation R is a well-ordering if it is a well-founded strict total order, as we did earlier, on page 61. By induction on the datatype, everything in it is a well-ordering (using proposition 60). The converse is a bit harder.

EXERCISE 109 *Prove the equivalence of these two definitions of well-ordering.*

Each of these two definitions can on its own justify a principle of induction over well-orderings. This principle takes two forms, one arising from each definition. Since the datatype of well-orderings is a rectype, we deduce an induction principle for it in an obvious way. On the other hand, there is a principle of well-founded induction that we can prove for each individual well-ordering, namely:

If $\mathfrak{X} = \langle X, <_X \rangle$ is a well-ordering, and P is a property such that $(\forall x \in X)(\forall y)(y <_X x \rightarrow P(y)) \rightarrow P(x))$, then $(\forall x \in X)(P(x))$.

DEFINITION 64

(i) $\hookrightarrow_{\mathfrak{X} \to \mathfrak{Y}}$ is to be the $\subseteq$-smallest bijection pairing the $<_X$-first member of X with the $<_Y$-first member of Y and closed under the following operation: if $X' \subseteq X$ and X' is mapped 1-1 onto $Y' \subseteq Y$ by $\hookrightarrow_{\mathfrak{X} \to \mathfrak{Y}}$, then $\hookrightarrow_{\mathfrak{X} \to \mathfrak{Y}}$ also pairs $x_{X'}$ with $y_{Y'}$, where $x_{X'}$ is the $<_X$-first element of $X \setminus X'$ and $y_{Y'}$ is the $<_Y$-first member of $Y \setminus Y'$.

(ii) *If* $\hookrightarrow_{\mathfrak{X}\to\mathfrak{Y}}$ *is defined on the whole of* X, *we write* $\mathfrak{X}\hookrightarrow\mathfrak{Y}$.

(iii) *If* $\hookrightarrow_{\mathfrak{X}\to\mathfrak{Y}}$ *is defined on the whole of* X *but is not onto* Y, *we write* $\mathfrak{X} \hookrightarrow \mathfrak{Y}$. *Thus* $\mathfrak{X} \hookrightarrow \mathfrak{Y}$ *iff* $\mathfrak{X}$ *is isomorphic to an initial segment of* $\mathfrak{Y}$.

DEFINITION 65 *A structure is rigid if it has no nontrivial automorphisms.*

THEOREM 66 *All well-orderings are rigid.*

Proof: Suppose $\mathfrak{X}$ is not rigid, and let x be the $<_X$-minimal member of X that is moved by an automorphism. So for some automorphism π we have $x < \pi(x)$. But then $\pi^{-1}(x) < x$, since π is an automorphism, and then x is not minimal.   ∎

COROLLARY 67 *Any isomorphism between two well-orderings* $\mathfrak{X}$ *and* $\mathfrak{Y}$ *is unique.*

LEMMA 68 $\hookrightarrow$ *is well-founded.*

Proof: Suppose A is a nonempty set of well-orderings such that no member of it injects into all the others. Let $\mathfrak{X} = \langle X, <_X \rangle$ be an arbitrary member of A. Since A has no element that injects into all others, there are at least some $\mathfrak{Y} = \langle Y, <_Y \rangle$ such that, when we construct the canonical injection $\hookrightarrow_{\mathfrak{Y}\to\mathfrak{X}}$ from $\mathfrak{Y}$ to $\mathfrak{X}$, there are bits of X that are not in the range of the canonical bijection. Let X' be the collection of elements x of X such that, for some $\langle Y, <_Y \rangle$, x is not in the range of the canonical injection $\hookrightarrow_{\mathfrak{Y}\to\mathfrak{X}}$.

We will show that X' has no least member under $<_X$. Suppose it does, and x is the $<_X$-least element of X'. Then, for some $\mathfrak{Y} \in A$, x is the first thing not in the range of $\hookrightarrow_{\mathfrak{Y}\to\mathfrak{X}}$. But then $\mathfrak{Y}$ injects into every well-ordering in A, contradicting the assumption that there is no such $\mathfrak{Y}$.   ∎

More graphically (because of the connexity of $\hookrightarrow$ (proposition 62)) every nonempty set X of well-orderings has a member that canonically injects into all members of X.

Now let us minute a few elementary facts about well-orderings.

EXERCISE 110 *If there is an order-preserving embedding* $\pi : X \to Y$, *then* $\langle X, <_X \rangle$ *canonically injects into* $\langle Y, <_Y \rangle$.

Exercise 110 actually characterises well-orderings in the sense that

EXERCISE 111 *A linear ordering* $\mathfrak{X}$ *is a well-ordering iff every linear order that can be embedded in* $\mathfrak{X}$ *is isomorphic to an initial segment of* $\mathfrak{X}$.

We saw a definition of ordinal exponentiation in definition 54. There is an alternative characterisation of ordinal exponentiation in terms of relational types of well-orderings. Let $\langle A, \leq_A \rangle$ and $\langle B, \leq_B \rangle$ be well-orderings of length α and β, respectively. Partially order the set of functions **with finite support**[1] from B to A by the **colex** ordering: $f < g$ iff at the **last** argument where they differ the value of f is less than the value of g.

EXERCISE 112

(i) *Check that this is indeed a well-ordering of the functions* $B \to A$ *with finite support and is of length* α^β.

(ii) *Give an example to show that, if we had ordered these functions by first difference instead of last, the result would not automatically be a well-order.*

This operation on well-orderings that corresponds to exponentiation is sometimes described as giving a **synthetic** treatment of exponentiation, and in the same sense concatenation and lexicographic product give "synthetic" treatments of addition and multiplication, respectively. But this is really to put the cart before the horse: it is well-orderings and operations on them that are conceptually prior to ordinals, not the other way round. In slogan form one might say:

(First-order) Ordinal arithmetic is the study of those relations between well-orderings for which $\simeq$ is a congruence relation.

The first nontrivial result we saw that involved exponentiation was Cantor's normal form theorem, theorem 58. It made us think about ordinals like ω^n and ω^ω. It would be nice to have natural examples of well-orderings of lengths other than ω. $\mathbb{N} \times \mathbb{N}$ ordered lexicographically is of length ω^2. And, in general, $\mathbb{N}^n$ ordered lexicographically is of length ω^n. We can well-order the set of all finite lists of natural numbers to a longer length than this by a variant of the lexicographic ordering, but the definition is forgettable because of complications that have to do with

---

[1] This means "on all but finitely many arguments the function takes value 0".

deciding how to compare lists of different lengths. In some ways a simpler way to present these ordinals is through well-orderings of polynomials by dominance. (See definition 37.) Consider the quadratics $\lambda x.(ax^2+bx+c)$ and order them by dominance. It is fairly clear that $\lambda x.(ax^2 + bx + c)$ is dominated by $\lambda x.(a'x^2 + b'x + c')$ iff $\langle a, b, c\rangle$ comes below $\langle a', b', c'\rangle$ in the lexicographic order of $\mathbb{N} \times \mathbb{N} \times \mathbb{N}$. So the set of quadratics, ordered by dominance, is of length ω^3. In fact, this holds for polynomials of higher degree as well, so the set of polynomials of degree n, ordered by dominance, is of length ω^{n+1}. Finally, the set of all polynomials (ordered by dominance) will be of order $\omega + \omega^2 + \omega^3 \cdots + \omega^n \cdots$. What is this ordinal? Well, $\omega^n + \omega^{n+1}$ is the same as ω^{n+1}, so it is simply the sup of all these ordinals, which, by definition, is ω^ω. Of course we could have got straight the definition of the well-ordering of finite sequences of natural numbers for another presentation of ω^ω, but the advantage of this version is that it can be easily upgraded. Consider now not the set of polynomials with coefficients in $\mathbb{N}$ but the much larger class of functions obtained by allowing exponentiation as well, so we can have expressions like

$$x^{x^x+x^3+x} + x^{x^{50}} + x^{200} + 137 \cdot x^3.$$

and consider what happens if we try to order *these* by dominance. The result is a well-ordering of length ϵ_0.

We must now return to the ideas of definition 61.

DEFINITION 69 *The cofinality of α, written '$cf(\alpha)$', is the least ordinal that is the length of a cofinal subsequence of something of length α.*

Thus $cf(\omega) = \omega$, and the cofinality of any successor ordinal is 1. The cofinality of ω^2 is ω because the limit points are a cofinal subsequence of length ω.

Notice that the relation 'f is cofinal in g' is *transitive*.

DEFINITION 70 *An ordinal α is **regular** if $\alpha = cf(\alpha)$; otherwise it is **singular**.*

Clearly cf is idempotent $(cf(cf(\alpha)) = cf(\alpha))$ because of transitivity of "cofinal in", so all cofinalities are regular.

I mentioned earlier the important triviality that every normal function has a fixed point. This is true because we can always obtain a fixed point by iterating ω times. This gives us fixed points of cofinality ω, which are typically singular. The assertion that normal functions have *regular* fixed points is a strong ("large cardinal") axiom.

EXERCISE 113 *Assuming that there are well-orderings whose carrier sets are uncountable, there will be a shortest one, up to canonical isomorphism. Its length is notated 'ω_1'. Prove that ω_1 is regular. You may use the axiom of countable choice.*

Without AC the only (infinite) ordinal we can prove to be regular is ω, though finding models of ZF where all infinite ordinals are singular is very difficult indeed. Come to think of it, you may well wonder how you can be sure that there are any uncountable ordinals in the first place – let alone *regular* uncountable ordinals. This is a consequence of a deeply mysterious theorem called **Hartogs' theorem**, which will be theorem 93, and states that for every set x there is a well-orderable set y that cannot be injected into x.

The fact that ω_1 is regular means that we cannot reach it by any countable iteration of a continuous function from On into itself: any operation that takes countable ordinals to countable ordinals can be iterated ω times: the sup is never ω_1 – because the way it is generated ensures that it is of cofinality ω!

7.2.1 Cardinals pertaining to ordinals

DEFINITION 71 *An* **initial** *ordinal is one such that the carrier set of any well-ordering of that length is larger than the carrier set of any well-ordering of any shorter length.*

Here, of course, by 'larger' we mean that there is an injection going one way but no injection (not just: no order-preserving injection) going the other way.

Assuming AC we can generalise exercise 113 to show that, for every ordinal α, the $\alpha + 1$th initial ordinal is regular. (This is standard set theory, but we will not need it.)

Although this definition relies on the conception of ordinals as isomorphism types of well-orderings, this will not cause much difficulty for the reader as the only initial ordinals we need to think about are – apart from the finite initial ordinals! – ω and ω_1, the first uncountable ordinal. (In the Von Neumann implementation of ordinal and cardinal arithmetic in ZFC, initial ordinals implement cardinals.) Ordinals below

ω are typically identified with natural numbers. The set of countable ordinals is sometimes called the **second number class**. This expression is Cantor's. (The *first* number class is of course $\mathbb{N}$!)

Assuming **full** AC (as is common in the study of well-founded sets) every cardinal corresponds to a unique initial ordinal. The $\alpha + 1$st (infinite) initial ordinal is ω_α ($\alpha + 1$st because we start counting at '0', so $\aleph_0$ is the first aleph. We **always** omit the subscript '0' in 'ω_0'!), and the corresponding cardinal number is $\aleph_\alpha$.

This notation makes sense even without AC. A cardinal of a well-orderable set is called an **aleph**, and the collection of alephs is naturally well-ordered. The αth aleph is notated '$\aleph_\alpha$'. Remember that 0 is the least element of $\mathbb{N}$, so the first aleph is $\aleph_0$!

The reader should beware of the vast difference between cardinal and ordinal exponentiation. ω^ω is a countable ordinal; $\aleph_0^{\aleph_0}$, like $\aleph_0$, is a cardinal, and it is bigger than $\aleph_0$. Confusion can be caused by the widespread reprehensible habit of writing 'ω' for '$\aleph_0$'.

Next we show

THEOREM 72 *Cofinalities are initial ordinals.*

Proof: Fix $\langle X, \leq_X \rangle$, a well-ordering of length ζ, with ζ regular. Suppose further that κ is the initial ordinal corresponding to ζ and that $\kappa < \zeta$. We will obtain a contradiction. We enumerate X (in a different order) as a κ-sequence: $\langle X, \leq_\kappa \rangle$. Delete from X any element that is $\leq_X$ something that is $\leq_\kappa$ of it. What is left is a subset of X cofinal in X in the sense of either ordering and which is of length κ at most, contradicting regularity of ζ.   ∎

So every regular ordinal is initial. Therefore, every countable ordinal $> \omega$ is singular and has smaller cofinality. This cofinality cannot be a smaller countable ordinal $> \omega$ because cofinality is idempotent. So

REMARK 73 *Every countable limit ordinal has cofinality ω.*

This generalises exercise 113. The successor ordinals of course have cofinality 1!   ∎

7.2.2 Exercises

EXERCISE 114 *Look at your answer to exercise 7 on page 30. What is the rank of the well-founded relation you discovered there?*

EXERCISE **115** *Use Cantor normal forms to show that every ordinal can be expressed as a sum of powers of 2.*

EXERCISE **116** *Verify that the class of well-orderings is closed under substructure and cartesian product.*

EXERCISE **117** *Verify that the end-extension relation between well-founded binary structures is well-founded.*

EXERCISE **118** *Verify that the transitive closure of a well-founded relation is well-founded.*

EXERCISE **119** *Complete the proof of the recursion theorem: theorem 3.*

EXERCISE **120** *Look again at exercise 19 on page 36 in chapter 3. You should now be able to do the following proof, which is slightly more standard. Turn G upside-down. It has a well-founded part (see exercise 20, which immediately follows exercise 19 on page 36). Use the recursion theorem to define a map from the well-founded part of G to {I, II}. Use the fact that all infinite plays are won by player II to show that one of the two players has a winning strategy.*

7.3 Rank

Our first encounter with ideas of well-foundedness were in connection with the idea of *termination*. When we find ourselves musing "In order to this I have first to do that, and in order to do that I must first do the other ... " we are reasoning about a *prerequisite* relation, and if the process we are contemplating is ever to halt, this prerequisite relation had better be well-founded. However, there is no reason to expect that the ancestral of the prequisite relation should be a total order. One task might spawn two subtasks, neither of which depends on the other. It is not hard to imagine that there could be a process whose prerequisite relation was something like that in the Hasse diagram of figure 7.1, where actions taken at stages further up the page rely on the successful completion of actions taken lower down the page on the same line. If the processes (the bottom points of the Hasse diagram) are all started simultaneously and run in parallel, then at stage ω we will be able to do the task located at point x.

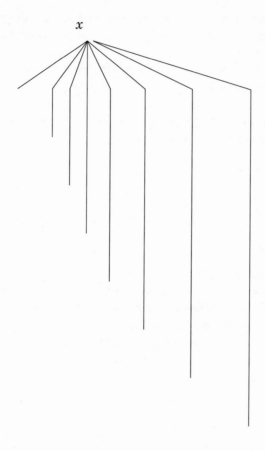

Fig. 7.1. A relation of rank ω

Suppose we have defined a function f by recursion on the relation whose Hasse diagram is in figure 7.1. How long does it take us to compute $f(x)$? Well, the recursion tells us to call f on all the immediate predecessors of x. These predecessors are nicely presented in such a way that computing the nth lets us in for a nest of subroutine calls of length n. Thus (assuming as we will that we can call simultaneously as many copies of this program as we like) after infinitely many steps we will have evaluated all of them and in one more step we will have computed $f(x)$. We would like to have some more information about what sort of infinity this is. Now look at the tree of figure 7.2.

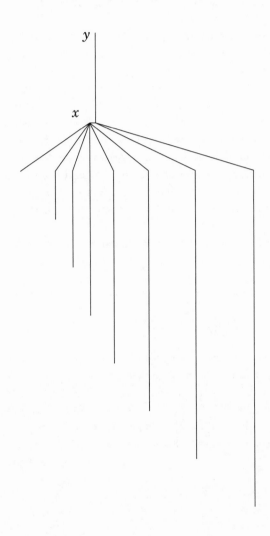

Fig. 7.2. A relation of rank $\omega + 1$

Suppose we have defined a function f by recursion on the relation R whose Hasse diagram is in figure 7.2. How long does it take to compute $f(y)$? Clearly it will take one step longer than it took us to compute $f(x)$, in the sense that we compute $f(y)$ *after* (one stage after) we compute $f(x)$.

What we are after is a parameter ("nastiness") associated with points x in the domain of R that tells us how hard it is to compute $f(x)$. Clearly, for these purposes, all R-minimal elements are equivalent and have nastiness 0. The nastiness of any element is at least as big as the nastinesses of its R-predecessors. Of course the complexity of the computation of $G(x, \emptyset)$ might well depend on x, but we are interested only in the contribution to the complexity made by R. Thereafter two points x and x' have the same nastiness as long as their R-predecessors are equally nasty. The thinking behind this is that, since we are also assuming unbounded parallelism, the *number* of R-ancestors of x has no effect on the nastiness of x: the only thing about them that matters is how nasty they are.

This function we have just defined is called **rank**, usually written with a 'ρ'.

DEFINITION 74 *If R is a well-founded relation, we define ρ by recursion on R:*

$$\rho(x) = sup\{(\rho(y)) + 1 : R(y, x)\}).$$

In fact, we have two natural ways to think of a rank function. We can either (as we have just done) define a rank function for each well-founded structure, so that it is a function that accepts elements of that structure and returns ordinals. The other thing we can do is associate with each structure the smallest ordinal that is not the rank (in the first sense) of any element of the structure. This we can think of as *the rank of the structure.*

EXERCISE 121 *Show that if $\langle X, R \rangle$ is a well-founded structure, then the rank of any point y in X is the same as the rank of $\langle {}^*R^{-1}\text{``}\{y\}, R|{}^*R^{-1}\text{``}\{y\}\rangle.$*

We are not going to make any use of ranks here, beyond pointing out that we have for any well-founded relation R a function from $dom(R)$ to the ordinals. What this means is that just as $\mathbb{N}$ has a privileged position among inductively defined sets (any induction over an inductively defined set with a first-order generating function can be thought of as an induction over $\mathbb{N}$), so $\langle On, \leq_{On}\rangle$ has a corresponding special position among well-founded relations: any induction over well-founded relations can be thought of as an induction on rank. Thus, instead of doing induction over some well-founded relation R to prove that everything in $dom(R)$ is ψ (where the induction hypothesis is "all R-predecessors of x are ψ" and we conclude that "x is ψ"), we prove by induction on rank that everything in $dom(R)$ is ψ (where the induction hypothesis is "everything of rank $< \alpha$ is ψ" and we conclude "everything of rank α is ψ").

THEOREM 75 *If $\mathfrak{M}$ is a rectype with carrier set M and with constructors of finite arity, then $\langle M, R\rangle$, where R is the engendering relation of $\mathfrak{M}$, is of rank precisely ω.*

Proof: This is because any element of an inductively defined structure, no matter how many founders or generating functions there are, is obtained by some finite number of applications of those functions to the founders. That is to say, we prove by structural induction on the structure that all its elements have finite rank. ∎

We cannot say much about the ranks of the engendering relations on rectypes that do not have finite character.

We have seen an application of ordinals as values of a parameter measuring lengths of computations with infinite parallelism ("nastiness"). This is not the only way in which people other than set theorists can naturally bump into them. Consider a computer system for storing sensitive information like people's credit information, or criminal records, and suchlike. It is clearly of interest to the subjects of these files to know who is retrieving this information (and when and why), and there do exist systems in which each file on an individual has a pointer to another file that contains a list of the the user IDs of people accessing the head file, and dates of those accesses. One can even imagine people wishing to know who has accessed *this* information, and maybe even a few steps further. A well-designed system would be able to allocate space for new and later members of this sequence of files as new reads by users made this necessary. These files naturally invite numerical subscripts. The system controllers might wish to know how many files had been generated by these reads, and know how rapidly new files were being generated, or what statistical relations existed between the number of reads at each level. This information would have to be stored in a file too, and the obvious subscript to give this file is ω. (It would not be sensible to label it '*n*', for *n* finite (even if large), because there is always in principle the possibility that we might generate *n* levels of data files.) Then we start all over again, with a file of user IDs and dates of people who have accessed the ωth file. Thus we can imagine a system where, *even though there are only finitely many nonempty files*, some of those files naturally have transfinite ordinals as subscripts.

Under any homomorphism of binary structures the image of a bad (no minimal elements) subset is also bad, so a homomorphic image of an ill-founded sructure is also ill-founded. This shows:

REMARK 76 *A binary structure is well-founded iff it admits a homomorphism onto a well-ordering.*

Do not worry about getting a formal proof of this until after we have understood the axiom scheme of replacement.

Related to this last result is the following exercise, rather in the spirit of chapter 3.

EXERCISE 122 *Every well-founded strict partial order can be refined to a well-ordering.*

It is just as well the result in this last exercise is true. If it were not, then the lecturer's linearisation problem might have unsolvable instances. As it is, once one has selected material for a course so that the *prerequisite* relation on it is well-founded, one can well-order what one wants to write. (The prerequisite relation might be a strict partial order, but we have to refine it to a strict total order because time is totally ordered.) If well-founded strict partial orders could not reliably be extended to well-orders, then well-foundedness of the prequisite relation would not be a sufficient condition for explainability! The fact that it is a *neccessary* condition for explainability bears a bit of reflection. Just think – all those areas of knowledge that will never be known because the prerequisite relation restricted to them is not well-founded!

EXERCISE 123 *Verify that the functions + and × defined by the rectype recursions correctly measure the lengths of the well-orderings given by concatenation and cartesian product.*

8

Set theory

Set theory is the first-order theory of equality and one extensional binary relation, and its importance in twentieth-century mathematics arises from the fact that any mathematical language can be interpreted in it, with varying felicitousness. In this respect it is a bit like graph theory. Graph theory is the theory of equality and one irreflexive symmetrical relation, and it is important because graphs are a useful data structure: things describable as graphs crop up all over the place. Set theory is more general still: not everything is a graph, but nobody has yet discovered a branch of mathematics that has successfully resisted formalisation into set theory. Indeed, that is the chief reason why so many mathematicians feel they have to know at least some set theory. The fact that everything can be expressed in set theory also causes it to be the natural site for the manifestation of foundational problems. That does not mean that (*pace* the American Mathematical Society's classification scheme, which has subject headings like "Logic and foundations") foundational problems are problems of set theory; it means merely that set theorists worry about them more than other people do. Many people (including a lot of set theorists) feel that the importance of set theory's status as the posessor of a universal language for mathematics has been exaggerated. This exaggeration has had the unfortunate consequence that many mathematicians feel that set theory is making claims to be more fundamental than the rest of mathematics and therefore in some sense more important. A related effect is that mathematicians tend to feel that set theorists are claiming to be able to deconstruct their discourse – and nobody likes having their discourse deconstructed. All this is a huge misunderstanding: set theory is a branch of mathematics like any other – except more fun.

8.1 Models of set theory

Set theory is just the theory of equality plus one extensional binary relation, so a model of set theory is just a set with an extensional relation on it. We noticed earlier that any binary relation E on a set X corresponds to a function $X \to \mathcal{P}(X)$ and that R is extensional iff this function is injective (see exercise 1(xix)). The Italian logician di Giorgi observed that this means that any structure at all for the language of set theory can be thought of simply as a collection A (of structureless atoms) with an injection i into $\mathcal{P}(A)$: simply associate with A the relation $\{\langle x, y \rangle : x \in i(y)\}$ to get a structure that looks like a toy set-theoretic universe. Accordingly, such maps are sometimes called 'di Giorgi maps'. However, for many purposes we want our models to have some extra properties as well. For a start, it is natural to ask that our set A of atoms might be a collection of *sets*. After all, elements of a model of set theory ought to be sets! So, although in principle we can think of A as a set of structureless atoms, and it is sometimes useful to do so, it is often more natural to think of A as a collection of *sets*.

If A is to be a collection of sets, it is natural to ask that i should somehow respect the membership structure that these sets happen to have. What does this actually amount to? The answer will come from considering a question that is primarily a philosophical question. If I am, as it were, *given* a set x, what else am I given (in the same sense)? With other kinds of entities this could be a real problem, but with sets the answer is plain. Since a set simply *is* the collection of its members, when we are "given" a set, we are "given" all its members and all *their* members, and so on, so we are "given" everything in $\bigcup_{n \in \mathbb{N}} \bigcup^n x$, which is the **transitive closure** of x. For the usual reasons this can also be characterised as the least transitive superset of x. (Now you understand why I prefer "ancestral" to "transitive closure" – it avoids overloading the word 'transitive'.)

In other words, if the elements of A are to be thought of as sets, and A is to be closed under the "if I am given x then I am given y" relation, then i must be the identity and A must be a transitive set. We will see later that this insistence will cost us little, as every (well-founded) model is isomorphic to one whose carrier set is a transitive set. See lemma 85.

If we restrict ourselves to using transitive models only, then something rather interesting happens. First a definition:

DEFINITION 77 $\ulcorner \phi^x \urcorner$ *is the result of replacing every quantifier* $\exists y$ *or* $\forall y$ *in* ϕ *by* $(\exists y \in x)$ *and* $(\forall y \in x)$. *'x' is assumed not to be free in* ϕ.

So ϕ^X is ϕ interpreted in X.

The reader might be wondering what difference there is between 'ϕ^X' and '$\langle X, \in \rangle \models \phi$'. The difference lies solely in the language in which the proposition is expressed. The first is an expression of the language of set theory, and the second is an expression of a metalanguage.

When do we have $\phi^X \longleftrightarrow \phi$? This question becomes interesting and natural once we restrict attention to transitive X's as we have agreed to. Quantifiers in the style $(\exists y \in x)$ and $(\forall y \in x)$ are **restricted**. If ϕ contains only restricted quantifiers, then, when we rewrite them in the manner required by definition 77, '$(\forall y \in x) \rightarrow \dots$' becomes '$(\forall y \in X \cap x)(\dots)$', but, because X was transitive and $x \in X$, we have $x \cap X = x$ as well, and the extra restriction has no effect, since all occurrences of 'X' vanish.

We will say that a formula containing only restricted quantifiers is Δ_0. What this discussion illustrates is the fact that if ϕ is Δ_0, then $\phi \longleftrightarrow \phi^X$ for any transitive X containing the denotations of the free variables in ϕ.

A typical set-theoretical axiom will say that the universe is closed under some operation or other. You might think that for a structure $\langle X, \in \rangle$ to be a model of such an axiom it would be sufficient for X to be closed under the operation in question. You would be wrong. The point is that the formula defining an operation may contain quantifiers, and if the range of those quantifiers is restricted to the carrier set of a structure, then the meaning of the formula may differ from the meaning it has when the quantified variables range over the whole universe. The power set axiom is an important example. If $\langle X, \in \rangle$ is to be a model of the axiom of power set, we require that, whenever $x \in X$, X should also contain the set $\{y \in X : (\forall w)(w \in X \rightarrow (w \in y \rightarrow w \in x))\}$ of all those elements of X all of whose members (that are in X) are also in x. Since we are assuming that X is transitive, this simplifies somewhat to $\{y \in X : (\forall w)(w \in y \rightarrow w \in x)\}$, which is just $\mathcal{P}(y) \cap X$. This might not be the same as $\mathcal{P}(x)$.

A Σ_{n+1} (resp. Π_{n+1}) formula is (a formula that is equivalent to) the result of binding with existential (resp. universal) quantifiers a Π_n (resp. Σ_n) formula (or Δ_0 if $n = 0$). A formula is Δ_n iff it is equivalent to both a Π_n formula and a Σ_n formula. The point of the power-set illustration is that "$y = \mathcal{P}(x)$" is Π_1 and not Δ_0.

Miniexercise: If a set x has some Σ_1 property in a universe M, it also has it in any end-extension of M. (We say that Σ_1 properties "generalise upwards".) Dually Π_1 sentences generalise downwards. Δ_0 properties are **absolute** (for all transitive structures).

I wrote at the beginning of section 7.2 about **end-extensions**. To each notion of limited quantifier there corresponds a notion of end-extension, and vice versa. As we saw there, in set theory the appropriate notion of end-extension is "new sets: yes; new members of old sets: no!". In general, whatever the binary relation involved, an end-extension is one that preserves formulæ without unlimited quantifiers: end-extensions are elementary (see definition 27) for sentences containing no unbounded quantifiers.

We shall say more about this when we encounter the axiom scheme of collection.

8.2 The paradoxes

In an ancient Greek puzzle, a man is walking by the banks of the Nile with his child. A crocodile jumps out of the river, seizes the child and says "I will release the child if you guess correctly whether or not I will return the child". The father replies "You will not return the child". The crocodile's reply is not recorded, but presumably it is so distracted by the intellectual puzzle that it drops the child. This does not sound very likely: there is no evidence that crocodiles are interested in logic. Indeed, I suspect that what distinguishes us not just from crocodiles but from other animals as well – even all other hominids – is not toolmaking or posession of language or any of the other skills that were believed at various times in the last century to be peculiar to us, but rather our capacity to lose sleep over logical paradoxes like the one that deprived the crocodile of its meal – which we may have been doing ever since the point at which we became *homo sapiens sapiens*. At any rate, written versions of these puzzles appear at the earliest possible stage in the record, at the point where people started using writing for thought and not merely (Linear A and Linear B) accountancy.

In Aristotle's *Rhetoric* there is the story of the lawyer Corax and his student Tisias. They agree that Tisias will pay Corax for his lessons when he wins his first case. Tisias then does not practice law, so Corax sues him for his fee. Tisias says: "Either I win this, in which case, by the judgement of the court, I do not have to pay Corax, or I lose. If I lose, then by our agreement I do not have to pay him, for I pay him only when I win my first case." Corax argues: "Either I win, in which case Tisias must pay me by the judgement of the court, or he wins, in which case he must pay me according to our agreement".

These puzzles could cause annoyance and perplexity in the ancient world, but with the spread of formalised mathematics to encompass logic about 100 ago they acquired the potential to cause real disaster as well.

Naïve set theory is the axiom of extensionality – the assumption that $\in$ is extensional – and the axiom scheme of naïve comprehension:

$$(\forall \vec{x})(\exists y)(\forall z)(z \in y \longleftrightarrow \Phi)$$

with 'y' not free in Φ. (Miniexercise: Why does this last clause matter?)

If we let Φ be '$z \notin z$', we obtain a contradiction immediately. This is merely the most immediate and straightforward of the paradoxes that arose at that time. The less straightforward paradoxes tend to partake of the obscure and semantical quality of the story of Tisias and Corax: Berry's paradox is the paradox of the smallest integer not definable in at most 19 syllables. Grelling's paradox is the paradox concerning the word "heterological". A word is *autological* if it is true of itself ("short", "english") and *heterological* if it isn't ("long", "German"). We get a paradox if we ask whether or not 'heterological' is heterological. The status of "autological" seems obscure too. This is a *semantic* paradox, not a *logical* paradox, in Ramsey's terms. This becomes apparent if one thinks about the word 'italicised', which makes it clear that we have to consider use – mention and type – token distinctions carefully. The paradox of the barber is, like Grelling's paradox, another presentation of Russell's paradox. In a certain village there lives a barber who shaves all those men (and only those men) who do not shave themselves. The usual answer is that of course (!) there is no such village. Another answer could be that the barber is a woman. Perhaps the best answer is that the barber lives outside the village.

Cantor's paradox arises from Cantor's theorem from section 2.1.6, which tells us that $|X| \not\geq^* |\mathcal{P}(X)|$. However, if we let X be V, the universe, we obtain a contradiction: the universe is its own power set, so the identity is a surjection from V onto its own power set.

Assume f is a bijection between X and $\mathcal{P}(X)$, and run the construction of the "diagonal" set: $\{x \in X : x \notin f(x)\}$. What is this object doing in the di Giorgi model? It is $\{x : x \notin x\}$. This object is the star in Russell's paradox, and this is how Russell discovered the paradox.

EXERCISE 124 *Show that the collection $\{x : \neg\exists y(x \in y \in x)\}$ cannot be a set.*

Show further that, for each n, the collection $\{x : x \notin^n x\}$ cannot be a set. (Russell's paradox is the case $n = 1$ and the paradox just mentioned is $n = 2$.) The reader may wish to look again at the characterisation of well-foundedness in terms of loops and descending chains on page 28.

These all seem to be the same. There is a '∞' version, known as **Mirimanoff's paradox**. This concerns the collection of all well-founded sets. The usual way to present this paradox uses the "wrong" definition of well-foundedness from page 29: R is well-founded if there is no sequence $\langle x_n : n \in \mathbb{N} \rangle$ of elements of the domain of R so that $\forall n\ R(x_{n+1}, x_n)$. We say that a *set* x is well-founded if there is no sequence $\langle x_n : n \in \mathbb{N} \rangle\ \forall n\ x_{n+1} \in x_n$ with $x_1 = x$. We obtain a paradox by asking whether the collection of well-founded sets is itself well-founded.

A more arresting way of presenting Mirimanoff's paradox is due to Bill Zwicker. Consider the collection of all games in which all plays are of finite length. (A game need not have a finite bound on the lengths of its plays to belong to this set). *Hypergame* is the following game. Player I picks a game of finite length, which I and II then procede to play, with II starting. A paradox arises if we ask whether or not Hypergame is a game of finite length. Related to this is the paradox at the end of `www.dpmms.cam.ac.uk/~tf/charlestalk.dvi`.

The last in our list of classical paradoxes is of course also the first: "I am lying". Prior (1976) has an interesting nonparadoxical version: Miniexercise: **"Everything I say is false"**. Why is it not paradoxical? What does it prove? For discussion of this see Prior (1976).

The paradoxes are normally thought of as products of the "crisis in foundations", when the growing tide of formalisation reached the diagonal arguments[1] and the resulting conflagration gave birth to axiomatised set theory, which we are about to see. But wilful disregard of the lessons of this period have enabled modern writers, too, to come up with paradoxes, some of which are quite instructive. Yablo's paradox is an interesting example of a modern paradox:

For each $i \in \mathbf{N}$, let S_i assert that for all $j > i$, S_j is untrue.

This makes a point about the ill-foundedness of the subformula relation. See the article at `www.dpmms.cam.ac.uk/~tf/yablondjfl.ps`.

---

[1] Notice the resemblance between the structure of the paradoxes and the proof of the unsolvability of the halting problem. My *Doktorvater* used to say that Euclid's proof of the infinitude of the set of primes was a diagonal argument.

8.3 Axioms for set theory with the axiom of foundation

This should really be subtitled *"Safe Sets"*.

The paradoxes in naïve set theory are intolerable, and if we are to use set theory, we will have to explicitly axiomatise it to get a system (or systems) that we can use without fear of contradiction.

A sensible approach is to start with things we trust and close under operations that preserve trustworthiness. With any luck the inductively defined collection that this produces will be rich enough to do all the work that we wish to heap on it. The declaration of the well-founded sets is as simple as can be:

DEFINITION 78 *The empty set is a well-founded set; every collection of well-founded sets is a well-founded set. Nothing else is a well-founded set.*

This gives us a formal-looking definition of the well-founded sets, namely,

DEFINITION 79 $WF := \bigcap \{Y : \mathcal{P}(Y) \subseteq Y\}$.

On the face of it, this definition ought to be vacuous: Cantor's theorem tells us that $\{Y : \mathcal{P}(Y) \subseteq Y\}$ is empty, and the intersection of the empty set is the universe. Clearly we are not going to be able to invoke this definition in a context where we have proved Cantor's theorem. We should be prepared for trouble because WF is a paradoxical object, as witness Mirimanoff's paradox. The problem lies in the arity of the constructor: set-of is not of finite arity, nor countable, nor of bounded arity at all. However, at this stage we are merely trying to find out what assumptions we have to make in order to give proper expression to the ideas we want to develop – and the idea of a rectype of inductively defined sets is definitely one we want to develop. While we are groping around we will use naïve set theory without worrying too much about the paradoxes, and keep on paying out thread like Ariadne so that when we get to somewhere interesting we can retrace our steps and discover what axioms we will need to get there safely in future.

When we do get a sensible set of axioms for WF we will of course find that – in the context they provide – this declaration of them will not work. Getting a workable definition of well-founded set, and proving induction over $\in$ for them, will occupy us extensively below. For the moment there are two basic facts that we will need frequently, and we had better have right at the outset a proof that they follow from this definition.

LEMMA 80 *If every member of x is well-founded, so is x.*

Proof: Suppose every member of x belongs to all X such that $\mathcal{P}(X) \subseteq X$. Then $x \subseteq X$ for all X such that $\mathcal{P}(X) \subseteq X$. Then $x \in \mathcal{P}(X)$ for all such X, whence $x \in X$ for all such X and x is well-founded as desired.∎

DEFINITION 81 *A set x is* **transitive** *if $\bigcup x \subseteq x$ or (equivalently) $x \subseteq \mathcal{P}(x)$.*

LEMMA 82 *WF is transitive: every member of a well-founded set is well-founded.*

If x is well-founded, and X is an arbitrary set satisfying $(\forall y)(y \subseteq X \rightarrow y \in X)$, then obviously $x \in X$. It will suffice to show that $x \subseteq X$ as well.

Suppose $x \not\subseteq X$. We will show that $\mathcal{P}(X) \setminus \{x\} \subseteq (X \setminus \{x\})$, whence $x \in (X \setminus \{x\})$ (since x is well-founded). This is impossible.

Suppose $y \subseteq (X \setminus \{x\})$. Then $y \subseteq X$ and $y \in X$. To deduce $y \in (X \setminus \{x\})$ it will suffice to show $y \neq x$, which would follow from $x \not\subseteq (X \setminus \{x\})$. But we have assumed that $x \not\subseteq X$, so a fortiori $x \not\subseteq (X \setminus \{x\})$.   ∎

Putting the last two results together we get

COROLLARY 83 *Every subset of a well-founded set is well-founded.* ∎

The collection of well-founded sets is the intersection of all Y such that $\mathcal{P}(Y) \subseteq Y$. This means that, if we have a property ϕ such that every collection of things that are ϕ is a set that is itself ϕ, then everything in the collection of well-founded sets is ϕ. That is to say, the following is a good rule of inference:

$$\frac{(\forall y)(y \in x \rightarrow \psi(y)) \rightarrow \psi(x)}{(\forall x \in WF)(\psi(x))}.$$

Now, if every set is well-founded (so $WF = V$), this simplifies to

$$\frac{(\forall y)(y \in x \rightarrow \psi(y)) \rightarrow \psi(x)}{(\forall x)(\phi(x))}.$$

This is ∈-induction.

We can also prove by ∈-induction that every set is a member of *WF*. This is pretty easy: take $\psi(x)$ to be "x is well-founded."

This shows:

THEOREM 84 ∈-*induction iff* $WF = V$.

The collection of well-founded sets is usually called the **cumulative hierarchy**.

8.4 Zermelo set theory

What axioms does this conception of set give rise to? To what axioms about sets will we be led by the assumption that every set is in WF?

- By lemma 80, any set of well-founded sets is well-founded, so the well-founded sets are a model for the axiom of **pairing**. This is $(\forall x)(\forall y)(\exists z)(\forall w)(w \in z \longleftrightarrow (w \in x \lor w \in y))$.

- By lemma 82, every member of a well-founded set is well-founded, so if x is well-founded, so is every member of it, and so is every member of $\bigcup x$. But a set is well-founded as long as all its members are, so the sumset of a well-founded set is well-founded. This gives us the axiom of **sumset**. This is $(\forall x)(\exists y)(\forall z)(z \in y \longleftrightarrow (\exists w)(z \in w \land w \in x))$.

- If every member of a well-founded set is well-founded, and every set of well-founded sets is well-founded, then any subset of a well-founded set is well-founded. This justifies **aussonderung** also known as **separation**. This axiom scheme is $(\forall x)(\forall \vec{w})(\exists y)(\forall z)(z \in y \longleftrightarrow (z \in x \land \phi(z, \vec{w}))$.

- In addition, the set of subsets of a well-founded set is a set of well-founded sets and is therefore a well-founded set itself by lemma 80. This justifies the axiom of **power set**, which is $(\forall x)(\exists y)(\forall z)(z \in y \longleftrightarrow z \subseteq x)$.

- Infinity? Well, if you keep on doing the construction, then after infinite time you will have built infinitely many well-founded sets, and at all stages thereafter the set of all of the things you have constructed so far will be an infinite well-founded set. So any sufficiently long initial segment of WF will be a model for a suitably phrased axiom of infinity.

- WF is a model of extensionality because it is transitive. If x and y are distinct well-founded sets, then $x \Delta y$ is well-founded and nonempty and there will be a well-founded member of it by lemma 80.

These are the axioms of **Zermelo set theory**. The only one I have not given explicitly is the axiom of infinity. This has various formulations, but for technical reasons that will be explained on page 180 it is usually given in the form: $(\exists x)(\emptyset \in x \land (\forall y \in x)(y \cup \{y\} \in x))$.

We arrived at the axioms of Zermelo set theory by asking what axioms ought to be true in the universe WF of well-founded sets. If we have followed this path correctly, we should now have a set of axioms that enable us to prove induction over $\in$ and the recursion theorem for $\in$, namely, that definitions by recursion over $\in$ are legitimate. The reason why this is not going to be completely straightforward is that the inductive definition we gave of "x is well-founded" only makes sense when the intersection of $\{X : \mathcal{P}(X) \subseteq X\}$ is nontrivial. Zermelo set theory proves Cantor's theorem, and it therefore proves that this intersection is trivial, and therefore that every set is well-founded in the inductive sense. We shall see later how, despite this, Zermelo set theory remains consistent if we add to it the assertion that there is an $x = \{x\}$. This obliges us to find other characterisations of well-founded sets that are not in danger of collapse into triviality in this way. There are three candidate definitions for well-foundedness in the context of the axioms we have so far, and we consider the justification of $\in$-induction for each of these three concepts of well-founded set.

They all involve the concept of what set theorists call the **transitive closure** of a set, $TC(x)$. This is the collection of all those things related to x by the transitive closure of $\in$.

They are

(i) "All descending $\in$-chains from x are finite" (every sequence $\{x_0, x_1, x_2 \ldots\}$, where $x_0 = x$ and for all i, $x_{i+1} \in x_i$, is finite).

(ii) Every subset of $TC(x)$ has an $\in$-minimal element.

(iii) x is nice: this idea is a special case of the R-regularity from theorem 2 when R is $\in$. A set x is **regular** iff

$$(\forall y)(x \in y \to (\exists w \in y)(w \cap y = \emptyset)).$$

Let us try to justify $\in$-induction for each of these in turn. In each case we will assume $(\forall y \in x)(\phi(y)) \to \phi(x)$ and let x be an arbitrary set that is well-founded in the sense-in-hand and $\neg\phi(x)$, and hope to deduce a contradiction.

(i) All descending $\in$-chains are finite

Every set that is not ϕ has a member that is not ϕ, so we can pick an *infinite* descending $\in$-chain starting at x and thereby deduce our desired contradiction. To do this properly we need DC, which readers will remember says that, if R is a relation such that $(\forall x \in Dom(R))(\exists y)(R(x,y))$, then there is an infinite R-chain. To use this axiom we seem to need to take R to be the $\in$-relation restricted to all the sets that are in x, or in something in x, or in something that is in something, and so on. That is to say, we seem to need $TC(\{x\})$ to be a set.

(ii) Every subset of $TC(x)$ has an $\in$-minimal element.

(Notice that this does not – on the face of it at least – commit us to having $TC(x)$ as a set.)

In this case we know that $\neg\phi(x)$, but every subset of $TC(x)$ has an $\in$-minimal element. We wish to deduce a contradiction. The obvious subset of $TC(x)$ to consider is $\{y \in TC(x) : \neg\phi(y)\}$, which has no $\in$-minimal element. The only way to get the existence of this set from our axioms seems to be to assume the existence of $TC(x)$ and use separation.

(iii) Regular sets.

Assume $TC(x)$ exists, and consider $\{z \in TC(\{x\}) : \neg\phi(z)\}$, which is a set by separation. x is a member of it, so it must be disjoint from one of its members. Suppose it is disjoint from a member w. Since $\neg\phi(w)$, w must have members that are also $\neg\phi$, and all these members will be in $TC(\{x\})$. We know too that none of them are $\neg\phi$ because they are in w, which is disjoint from $\{z \in TC(\{x\}) : \neg\phi(z)\}$. So $\phi(w)$ too. Contradiction.

Again we seem to have had no choice but to use an axiom giving us the existence of transitive closures.

8.5 *ZF* from Zermelo: replacement, collection and limitation of size

The trouble with Zermelo is that, if we try to write out a formal proof of the recursion theorem for any of these three concepts of well-founded set we find that we need the axiom of **transitive containment** ("every set has a transitive superset"), and this is not an axiom of Zermelo. We could just adopt transitive containment as an axiom, but if we are prepared to be patient, we can motivate a stronger axiom that implies it.

The rectype WF does not have finite character, and although that is not *obviously* a *prima facie* problem, it is a problem in this case, because there is a theorem waiting in the wings ready to give us trouble should we ever wish to pretend that the creation of WF ever gets completed. Cantor's theorem tells us that no set can be equal to its power set, and a completed WF would certainly be a set equal to its own power set. So if we believe Cantor's theorem, we are never going to trust the *whole* of WF, but only some fragment of it. What can we say about the fragment that we trust? What operations is it closed under, for example?

We can ask this question in general about any rectype that lacks finite character and whose completion might therefore be problematic. How much do we trust? One thing is clear: the part that we trust is certainly going to be downward-closed under the engendering relation: after all, *prima facie* the only reason for trusting a set is trust in the things that engender it.

In the case of WF this tells us that, if we think that a set exists, we must also think that all its members exist, and so on. In short, we must think that everything in its transitive closure exists. Does this mean that if we trust x, we should also trust $TC(x)$? The answer to that depends on how cautious a strategy we want. Are there any other considerations that make trusting $TC(x)$ for trustworthy x sound like a sensible thing to do? Well, one of them we have just seen, namely, that if we believe that $TC(x)$ exists for all x, then we can justify $\in$-induction for well-founded sets. But we can say more than this.

The "limitation of size" principle says that **anything the same size as a set is a set**, or alternatively, **anything that isn't too big is a set**. If we think of this as a way of avoiding the paradoxes, this is completely barmy: whether or not a set is paradoxical seems to have much more to do with how kinky its definition is than with how big it is. In addition, unless we assume the axiom of foundation *ab initio* it is perfectly clear that not everything the same size as a well-founded set is well-founded. If $x = \{x\}$, this x is the same size as any other singleton, but it is not well-founded. However, the idea that WF is a model for replacement is not obviously barmy. Here is the axiom scheme of replacement:

If $(\forall x)(\exists! y)(\phi(x,y))$, then $(\forall X)(\exists Y)(\forall z)(z \in Y \longleftrightarrow (\exists w \in X)\phi(w,z))$ (ϕ represents a function, and replacement says "the image of a set in a function is a set").

Now it is quite clear that WF is indeed a model for replacement, since the image of a well-founded set in a function that takes well-founded sets to well-founded sets will be well-founded. However, whether we believe that the trustworthy initial segment of WF is closed under replacement is a matter of taste. **Zermelo-Fraenkel** set theory is Zermelo set theory with the axiom scheme of replacement.

Replacement immediately gives us the existence of transitive closure s. Fix a set X, and consider the recursively defined function f that sends 0 to X, and sends $n+1$ to $\bigcup(f(n))$. This is defined on everything in $\mathbb{N}$. By replacement its range is a set. We then use the axiom of sumset to get the sumset of the range, which is of course $TC(X)$.

This does look a bit dodgy: what exactly is the ϕ in the instance of replacement we are using here? It is to an explanation of this that we now turn.

8.5.0.1 *Bootstrapping the recursion theorem in ZF*

In order to prove the recursion theorem for $\in$ along any of the lines above we need to know that transitive closures exist. Now the obvious way to exploit replacement to obtain the transitive closure of x is to apply $\lambda n.\bigcup^n x$ to $\mathbb{N}$ to form the set $\{x, \bigcup x \ldots \bigcup^n x \ldots\}$ and take its sumset. This uses the recursion theorem, so we must find a way of getting the transitive closure from the other axioms without using the recursion theorem.

Clearly we are not going to be able to just magic $\{x, \bigcup x \ldots \bigcup^n x \ldots\}$ into existence as the intersection of all sets containing A and closed under $\bigcup$, since we have not got an axiom giving us a set containing x and closed under $\bigcup$. And how are we going to define the obvious bijection between $\{x, \bigcup x \ldots \bigcup^n x \ldots\}$ and $\mathbb{N}$? This looks like an inductively defined set again and we are back where we started.

The simplest way to deal with this is the concept of a partial map satisfying the recursion wherever it can, usually in the slang called an **attempt**. (We first encountered this idea in the proof of theorem 3.) We prove by induction on the naturals that for all n there is a function defined on the naturals up to n that satisfies the recursion *and that this function – or at least the restriction of any such function to the naturals below n – is unique.* The ϕ we want is the formula that says that y and n are related iff every attempt defined at n sends n to y.

Conway's principle, in the appendix to the first ("zero-th") part of Conway (2001), is that all recursive datatypes exist. Well, the example of WF shows it fails for certain rectypes of unbounded arity, but the idea is not a bad one. Any axiomatic set theory that is to be a serious candidate for a system into which we can implement the whole of mathematics should give a smooth, unified and intelligible proof of those instances of

Conway's principle that deserve proofs. How does ZF shape up to this challenge?

The naïve way to attempt to prove the existence of the rectype containing founders in F and closed under constructors in C is to take the collection X of all sets containing all of F and closed under all constructors in C, and then take its intersection. This naïve strategy does not work because except in trivial cases X cannot be a set. The trouble is that, for some small n, $\bigcup^n X$ turns out to be the universe.

However, help is at hand, in the form of exercise 36, which the reader should now review. It tells us that, in order to construct the rectype containing founders in F and closed under constructors in C, it is not necessary for the collection X of all sets containing all of F and closed under all constructors in C to be a set. It is sufficient for there to be even one set $Y \supset F$ that is closed under constructors in C, for everything in the desired rectype will be in Y and so the rectype will be a set by separation.

So far so good, but how are we to come by such a set Y, given F and C? Let us start by considering rectypes of finite character. We saw in section 2.1.6 that every rectype with finitely many founders and finitely many operations all of finite arity is a countable set. We can now explain how to give a more formal proof of this fact, using the axiom of infinity. The axiom of infinity has many forms, but the commonest is one that states that there is a set X that contains $\emptyset$ and is closed under $\lambda x.(x \cup \{x\})$. As we shall see, $\emptyset$ is 0 in the von Neumann implementation of ordinals, and $\lambda x.(x \cup \{x\})$ is successor in the same implementation. This is the give-away that that form of the axiom of infinity was set up precisely so that we can apply to it the trick of exercise 36 to obtain the von Neumann implementation of $\mathbb{N}$.

We can then construct any other rectype of finite character by means of replacement. The idea is that any finitely presented rectype will be a surjective image of $\mathbb{N}$ and therefore a set. How do we find a surjection? By means of the prime powers trick. Any element of a finitely presented rectype has a finite description – its *proof* in the sense of the discussion on page 43. A proof is a finite string of characters from a finite alphabet and therefore can be coded as a natural number by means of the prime powers trick of section 2.1.6.1. The decoding map from $\mathbb{N}$ to the rectype is therefore a surjection and the rectype will be a set.

8.5.1 Mostowski

Now that we have the axiom scheme of replacement we have transitive closures of sets and can prove the recursion theorem formally.

LEMMA 85 *(Mostowski's collapse lemma)*

(i) *If $\langle X, R \rangle$ is a well-founded extensional structure, then there is a **unique** transitive set Y and a unique isomorphism between $\langle X, R \rangle$ and $\langle Y, \in \rangle$.*

(ii) *If $\langle X, R \rangle$ is a well-founded structure, then there is a transitive set Y and a homomorphism $f : \langle X, R \rangle \rightarrow \langle Y, \in \rangle$.*

Proof: We use the recursion theorem. Set $\pi(x) := \{\pi(y) : yRx\}$. The desired Y is simply the range of π. Y is transitive because nothing ever gets put into Y unless all its members have been put in first. If R is extensional, then no two things in X have the same set of R-predecessors and so no two things ever get sent to the same thing by π.

8.6 Implementing the rest of mathematics

If we are to consider how to implement the rest of mathematics in set theory we will not be able to put off any longer a decision about how to implement ordered pairs. Ordered pairs of sets are not *prima facie* sets, but if we are to try to pretend that every mathematical object is a set, we will have to find a way of coding ordered pairs of sets as sets. It is easy to check that we can implement $\langle x, y \rangle$ as $\{\{x\}, \{x, y\}\}$, and this is, if not actually universal practice, so nearly universal that most readers will never encounter any other way of doing it. However, it is very important to know that other ways are available. The following exercise spells out this point:

EXERCISE 125 *Suppose f and g are two functions defined on the whole of the universe, with the ranges of f and g complementary (so that everything is value of f or of g but not both). Use f and g to define an ordered pair function.*

8.6.1 Scott's trick

The obvious way to implement ordinals is to take them to be isomorphism classes of well-orderings. Obvious it may be, but sadly it does not work as long as we have sumset and separation. Consider the ordinal number 1. This would be the set of all well-orderings of length 1. A well-ordering is the ordered pair of a set X and a relation $R \subset X \times X$ that well-orders X. The only well-ordering of a singleton is the empty relation, so the ordinal 1 would be the set of all ordered pairs $\langle \{x\}, \emptyset \rangle$, which is to say the set of all sets of the form $\{\{\{x\}\}, \{\{x\}, \emptyset\}\}$. So $\bigcup 1$ would be the set of all sets of the form $\{\{x\}\}$ or $\{\{x\}, \emptyset\}$, and so on, so that $\bigcup^3 1$ would be the universe. Separation then gives us the Russell set.

This problem is quite general in ZF: no mathematical object that one naturally thinks of as an isomorphism class can ever be a set. However, as long as one has the axiom of foundation, one can exploit the following trick, due to Dana Scott.

For each well-ordering, there will be a first stage in the cumulative hierarchy at which a well-ordering of that length appears. So we take the ordinal of that well-ordering to be the set of well-orderings isomorphic to it that appear at that stage in the cumulative hierarchy. This is a set by separation, and it will do very well. And, naturally, the same idea will work for any other mathematical object arising naturally as an isomorphism class.

Although this is a useful general idea – and we will use it – it is not actually the ideal way to implement ordinals in ZF. The implementation of ordinals that is universally used in ZF (so universally used that many set theorists think that they *are* ordinals) is due to von Neumann.

8.6.2 Von Neumann ordinals

We must cast our minds back to the characterisation of ordinal arithmetic as *that part of set theory for which isomorphism of well-orderings is a congruence relation*. The naïve thing is to take ordinals to be the equivalence classes. Sadly, as we have just seen, that does not work. We can use Scott's trick, and implement ordinals so that the ordinal of a well-ordering $\langle X, < \rangle$ is the set of all well-orderings of minimal rank isomorphic to $\langle X, < \rangle$. However, we can do something nicer. We know by Mostowski's collapse lemma (lemma 85) that every well-ordering is isomorphic to a unique transitive set well-ordered by $\in$, so each equivalence class will contain a unique well-ordering $\langle X, \in \rangle$, whose ordering relation is $\in$, so we can take these special well-orderings as our representatives. Next we notice that of our special representatives no two have the same carrier set. So we can discard the orderings on the carrier sets and just keep the carrier sets. That is to say we can take ordinals to be transitive sets well-ordered by $\in$. This is the **von Neumann** implementation of ordinals.

I write of the von Neumann *implementation* of ordinals and not the von Neumann *definition* of ordinals for two reasons. First, it is philosophically correct: we are not attempting to *define* ordinals, we are merely trying to find sets naturally equipped with relations and operations that mimic them. Von Neumann "ordinals" are not ordinals (Quine used to call them *counter sets*). Second, it is rhetorically prudent: it is important not to annoy other mathematicians by exaggerating what set theory can do for foundations.

The other way we can arrive at the von Neumann implementation is to think of ordinals as a rectype. Take 0 to be the empty set; take $\mathsf{succ}(\alpha)$ to be $\alpha \cup \{\alpha\}$ and sup to be $\bigcup$.

8.6.2.1 The cumulative hierarchy again

We can now define the cumulative hierarchy as the range of a function defined on the ordinals by means of the recursion theorem. Define

$$V_0 := \emptyset; \quad V_{\alpha+1} := \mathcal{P}(V_\alpha); \quad V_\lambda := \bigcup_{\beta < \lambda} V_\beta.$$

WF is a well-founded structure and therefore has a rank function: Mostowski's collapse lemma (lemma 85) will enable us to show that definition 74 is legitimate. Then we prove a connection between these two: by induction, the rank of a set is the least α s.t. it is in V_α.

8.6.3 Collection

The **axiom scheme of collection** states: $(\forall x \in X)(\exists y)(\psi(x,y)) \rightarrow (\exists Y)(\forall x \in X)(\exists y \in Y)(\psi(x,y))$, where ψ is any formula, with or without parameters.

Weaker versions of collection (e.g., for ψ with only one unrestricted quantifier) are often used in fragments of ZFC engineered for studying particular phenomena.

THEOREM 86 *WF* $\models$ *Collection and replacement are equivalent.*

Proof: Collection $\rightarrow$ replacement is easy. To show that replacement implies collection, assume replacement and the antecedent of collection, and derive the conclusion. Thus

$$(\forall x \in X)(\exists y)(\psi(x,y)).$$

Let $\phi(x,y)$ say that y is the set of all z such that $\psi(x,z)$ and z is of minimal rank. (Scott's trick put to use already!) Clearly ϕ is single-valued, so we can invoke replacement. The Y we want as witness to the "$\exists Y$" in collection is the sumset of the Y given us by replacement.   ∎

8.6.3.1 Quantifier pushing

We have encountered restricted quantifiers already (see page 76, for example) and we saw a hierarchy of classes of formulæ. There is a **hierarchy theorem** about this collection, and it has several parts. One part claims that every formula belongs to one of the classes Σ_n and Π_n, and the second part claims that the classes are all distinct. The second part is in severe danger if $V \neq WF$: when there is a universal set, any formula ψ is equivalent to both $(\exists x)(\forall y)(y \in x \land \psi^x)$ and to $\forall x \exists y (y \notin x \lor \psi^x)$. This is a crude fact, but the question of whether or not $V = WF$ has some subtle implications for the first part too.

For the first part, it is a miniexercise to show that all the Π^n and Σ_n classes are closed under conjunction and disjunction (or at least that their closures under logical equivalence are so closed). It is not at all obvious that they (or their logical closures) are closed under restricted quantification. Would we expect them to be? There are sound philosophical reasons why we might – at least if $V = WF$. In general, if we are dealing with a rectype, and if the restricted quantifiers are restricted in the style '$(\exists x)(R(x,y) \land \ldots)$' and '$(\forall x)(R(x,y) \to \ldots)$', where R is the engendering relation, then Δ_0 formulæ behave in many ways as if they contained no quantifiers at all. An unrestricted quantifier is an injunction to scour the whole universe in a search for a witness or a counterexample; a restricted quantifier invites us only to scour that part of the universe that lies in some sense "inside" something already given. The search is therefore "local" and should behave quite differently: that is to say, restricted universal quantification ought to behave like a finite conjunction and ought to distribute over disjunction in the approved de Morgan way. We saw how restricted quantifiers in the arithmetic of ℕ behave well in exercise 76.

What we will now see is that, if we have the axiom scheme of collection, then we can prove an analogue of the prenex normal form theorem:

THEOREM 87 *Given a theory T, which proves collection, for every expression ϕ of the language of set theory there is an expression ϕ' s.t.*

$T \vdash \phi \longleftrightarrow \phi'$ *and every restricted quantifier and every atomic formula occurs within the scope of all the unrestricted quantifiers.*

Proof: It is simple to check that $(\forall x)(\forall y \in z)\phi$ is the same as $(\forall y \in z)(\forall x)\phi$ (and similarly $\exists$), so the only hard work involved in the proof is in showing that

$$(\forall y \in z)(\exists x)\phi$$

is equivalent to something that has its existential quantifier out at the front. (This case is known in logicians' slang as "quantifier pushing".) But by collection we infer

$$(\exists X)(\forall y \in z)(\exists x \in X)\phi,$$

and the implication in the other direction is easy. The remaining case is where we have an unrestricted $\forall$ within the scope of a restricted $\exists$. But this case is subsumed under the one we have just dealt with, since it is its negation. After all, if $p \longleftrightarrow q$, then $\neg p \longleftrightarrow \neg q$. ∎

So the argument for replacement is that it enables us to prove the hierarchy theorem for the theory of well-founded sets, which ought to be provable, and which we do not seem to be able to prove otherwise.

8.6.4 Reflection

If $\phi \longleftrightarrow (\phi^{V_\gamma})$, we say γ **reflects** ϕ.

Unless ϕ is Δ_0, there is no reason to expect that there are any γ that reflect ϕ. The **reflection principle** says that there is nevertheless always such a γ. I shall prove something apparently slightly stronger than this.

THEOREM 88 *For every ϕ, ZF proves $\phi \longleftrightarrow (\exists$ a closed unbounded class of $\alpha)\phi^{V_\alpha}$.*

Proof: By induction on quantifiers and connectives. It is certainly true for ϕ a Δ_0 formula. Assume $\forall \vec{x} \exists \vec{y} \phi$. Then, in particular, for any old ordinal α, $(\forall \vec{x} \in V_\alpha)(\exists \vec{y})\phi$. Now by collection we infer $(\exists B)(\forall \vec{x} \in V_\alpha)(\exists \vec{y} \in B)\phi$, and we can take this B to be a V_β, getting $(\exists \beta)(\forall \vec{x} \in V_\alpha)(\exists \vec{y} \in V_\beta)\phi$, so we have proved $(\forall \alpha)(\exists \beta)(\forall \vec{x} \in V_\alpha)(\exists \vec{y} \in V_\beta)\phi$. That is to say, we have proved that the function $\lambda\alpha.(\text{least } \beta)(\forall \vec{x} \in V_\alpha)(\exists \vec{y} \in V_\beta)\phi)$ is total.

By induction hypothesis there is a closed unbounded class of ordinals that reflect ϕ. Let X be such a class of ordinals and consider the function

$\lambda\alpha.(\text{least }\beta \in X)(\forall \vec{x} \in V_\alpha)(\exists \vec{y} \in V_\beta)\phi)$. Since $\beta \in X$, this becomes $\lambda\alpha.(\text{least }\beta \in X)(\forall \vec{x} \in V_\alpha)(\exists \vec{y} \in V_\beta)\phi^{V_\beta})$. This function (or rather its restriction to X) is a continuous function from X into itself and will have a closed unbounded class of fixed points (by exercise 3 in section 3.1.3 on page 55). A closed unbounded subclass of a closed unbounded class is itself closed and unbounded. The ordinals in this class reflect $\forall \vec{x} \exists \vec{y} \phi$.   ∎

The theorem scheme of reflection is a kind of omnibus existence theorem for recursive datatypes, since it tells us that, if the universe is closed under a bundle of operations, then there is a set (indeed, many sets) closed under those same operations.

We needed the function $\lambda\alpha.(\text{least }\beta \in X)(\forall \vec{x} \in V_\alpha)(\exists \vec{y} \in V_\beta)\phi^{V_\beta})$ to be continuous so we could be sure that it had a closed unbounded class of fixed points. It is continuous because of the finitary nature of ϕ. If we were trying to prove reflection for an infinitary language (of the kind where we can bind infinitely many variables simultaneously), then the function would not be continuous. Thinking about what one can retrieve in this situation gets one into the kind of mental habits that prepare one for large cardinal axioms. As they say on TV: *do not attempt this at home.*

It is sometimes convenient to accord a kind of shadowy existence to collections that are not sets, particularly if there are obvious intensions of which they would be the extensions – like the collection of all singletons, or all things that are equal to themselves (the intensions are pretty straightforward after all!). We call these things **classes** or (since some people want to call all collections "classes" – so that sets are a kind of class) **proper classes**.

If we allow classes, we can reformulate ZF as follows. Add to the language of set theory a suite of uppercase Roman variables to range over classes as well as sets. Lowercase variables will continue to range solely over sets, as before. Since classes are sets that are not members of anything, we can express "X is a set" in this language as '$(\exists Y)(X \in Y)$' and we do not need a new predicate letter to capture sethood.

Next we add an axiom scheme of class existence: for any expression $\phi(x, \vec{y})$ whatever, we have a class of all x such that $\phi(x, \vec{y})$.

We rewrite all the axioms of ZF except replacement and separation by resticting all quantifiers to range over sets and not classes. We can now reduce these two schemes to single axioms that say "the image of a set in a class is a set" and "the intersection of a set and a class is a set". Does this make for a finite set of axioms? This depends on whether the axiom scheme of class existence can be deduced from finitely many instances of itself. The version of this scheme asserted in the last paragraph cannot be reduced to finitely many instances. This system is commonly known as Morse-Kelley set theory.[1] However, if we restrict it so that for ϕ to appear in a class existence axiom it must not have any bound class variables, then it can be reduced to finitely many axioms, and this system is usally known as 'GB' (Gödel-Bernays). GB is exactly as strong as ZF, in the sense that for some sensible proof systems at least there is an algorithm that transforms GB proofs of assertions about sets into ZF proofs of those same assertions. Indeed, for a suitable gnumbering of proofs, the function involved is primitive recursive.

Morse-Kelley is actually stronger than GB, and although the details are hard, it is not hard to see why this might be true. Because Morse-Kelley allows bound class variables to appear in the set existence axioms, it proves the existence of more sets and therefore makes it possible to prove more things by induction. That was the the point of the discussion in section 2.1.5.

COROLLARY 89 *ZF is not finitely axiomatisable.*

Proof: If ZF were finitely axiomatisable, then by reflection there would be an ordinal α such that $\langle V_\alpha, \in \rangle$ were a model of ZF. This V_α is a set. This is important because, once we have a gnumbering of formulæ, the assertion that every formula in some semidecidable set Σ of formulæ is true in $\langle V_\omega, \in \rangle$ is an expression in the language of set theory, and we can set about proving that all logical consequences of ϕ are also true in $\langle V_\alpha, \in \rangle$. We do this by structural induction on proofs. Then we will have established that the set of logical consequences of Σ has a model and is free of contradiction. We know because of Gödel's incompleteness theorem that no theory can prove its own consistency, so no initial segment $\langle V_\alpha, \in \rangle$ can be a model of ZF. Reflection tells us that if ZF were finitely axiomatisable, we would be able to find such an initial segment. So ZF is not finitely axiomatisable. ∎

Attempts to construct consistency proofs for a theory T by proving by structural induction on proofs that all its consequences are true in some structure $\langle X, R \rangle$ where at least one of X and R is a proper class run the risk that the induction will fail because the proposition being

---

[1] It was actually first spelled out by Wang (1949).

proved by the induction contains allusions to classes in a way that will sabotage the induction. We saw earlier how the success of an induction can depend sensitively on existence theorems for sets, in that the collection of putative counterexamples must be a set. Models whose carrier sets or relations are proper classes cannot be relied upon to support the inductions on which consistency proofs rely.

In fact, we can show something slightly stronger than corollary 89: *ZF* proves the consistency of any of its finitely axiomatisable subsystems. If ϕ is the conjunction of all the axioms of a finite fragment of *ZF*, we have $ZF \vdash \phi$, so for some β, $V_\beta \models \phi$.

EXERCISE 126 *Von Neumann had an axiom that makes sense in the context of set theory with classes:*

A class is a set iff it is not the same size as V.

Prove that von Neumann's axiom is equivalent to replacement plus choice.

Finally some important trivialities:

Replacement gives bigger sets

Using replacement we can prove the existence of the set $\{\mathbb{N}, \mathcal{P}(\mathbb{N}), \mathcal{P}^2(\mathbb{N})\dots\}$ and then its sumset, which is of course bigger than $\mathcal{P}^n(\mathbb{N})$ for any $n \in \mathbb{N}$. What is perhaps slightly more surprising is that replacement enables us to prove novel results, not provable in Zermelo, about small sets whose existence even Zermelo set theory can prove. Thus there are theorems of analysis provable in ZF that are not provable in Zermelo. The most celebrated of these is D. A. Martin's proof of a generalisation of exercise 19 to the effect that games where I and II pick natural numbers, and the set of plays that give rise to wins for I is Borel, then there is a winning strategy for one player or the other. (Martin 1975).

The philosophically motivated reader may have been worried by the cheerful and casual way in which we adopted the axiom scheme of replacement. It is true that it implies the existence of transitive closures and thereby gives us a proof of the recursion theorem, but its philosophical motivation is weak. On the other hand, it does also imply the normal form theorem for restricted quantifiers (theorem 87) and reflection, both of which are philosophically appealing in a way that replacement itself is not. It also proves lemma 85, and although it will not be apparent yet, lemma 85 is not merely appealing but actually indispensible. The fact that the axiom scheme of replacement implies it is a very powerful point in its favour. *By their deeds ye shall know them*, and in the end one has to judge an axiom by its consequences. This probably sounds to the reader like an instance of the fallacy of affirming the consequent, but this is actually quite legitimate: pointing out that a candidate axiom gives a single reason for believing a lot of things that we have disparate reasons for wishing to believe is a very good way of arguing for it. It is an example of Occam's razor. I was attracted to Buddhism because it seemed to give a single reason for being atheist, vegetarian and pacifist, all of which I was anyway. A single explanation for a lot of hitherto apparently unconnected phenomena is *prima facie* more attractive than lots of separate explanations: that is the appeal of conspiracy theories.

8.7 Some elementary cardinal arithmetic

If there is a bijection between X and Y, we say that X and Y are **equinumerous** or **equipollent**. Equipollence/equinumerosity is obviously an equivalence relation, and cardinal arithmetic is that part of set theory for which it is a congruence relation. Cardinal arithmetic is not a branch of set theory, but it can be implemented in set theory by means of Scott's trick. However, if we assume AC, then every set can be well-ordered, so we can take the cardinal of a set simply to be the unique initial ordinal corresponding to it. This is normal practice among set theorists, but if we leap into it too quickly, we miss the chance of drawing some morals from what we have done so far. The Schröder-Bernstein theorem (theorem 3.1.1) says that $\leq$ on cardinals is antisymmetric. This is a nontrivial fact and should not be taken for granted. It is true for cardinals (and for ordinals, where it is very easy to prove), but it is not true for arbitrary linear order types, for example, ($[0, 1)$ and $(0, 1]$ embed in each other but are not isomorphic).

We have not yet defined addition, multiplication or exponentiation for cardinals, but the definitions are obvious: whenever $\alpha = |A|$ and $\beta = |B|$ we can set $\alpha \cdot \beta = |A \times B|$ and $\alpha^\beta = |A \to B|$. If, additionally $A \cap B = \emptyset$, then $\alpha + \beta = |A \cup B|$. There is an obvious bijection between the disjoint union $A \sqcup B$, and the disjoint union $B \sqcup A$, and it is in virtue of this that cardinal addition is commutative. Similarly, there is an obvious bijection between $A \times B$ and $B \times A$ in virtue of which cardinal multiplication is commutative. (It is in virtue of the same bijection – through the Curry-Howard correspondence – that $A \wedge B \longleftrightarrow B \wedge A$!)

EXERCISE 127 *There are distributive laws for addition, multiplica-tion and exponentiation of cardinals. State them correctly, and indi-cate what assertions of propositional logic correspond to them under the Curry-Howard correspondence. (Hint: Disjunction corresponds to disjoint union.)*

But cardinal arithmetic has some nontrivial theorems as well.

PROPOSITION 90 *If x is an infinite set that can be well-ordered, $|x|^2 = |x|$.*

Proof: Here is an obvious strategy. It does not work properly, but the idea is a good one, and it will lead us to a strategy that does. Let us overload '$|$' by writing '$|\zeta|$' for the cardinal number of a set that admits a well-ordering of length ζ. Notice that the equivalence relation $|\alpha| = |\beta|$ is a congruence relation for all the operations of ordinal arithmetic. Use this to prove by induction on α that $(\forall\alpha)(|\alpha| = |\alpha^2|)$. The induction step at successor ordinals seems fine: if $|\alpha| = |\alpha^2|$, then $|(\alpha+1)^2| = |\alpha^2+\alpha+1|$ (at least if $\alpha > \omega$, so that $1 + \alpha = \alpha$.) Then $|\alpha^2 + \alpha + 1| = |\alpha + \alpha + 1|$ and clearly (again, as long as $\alpha > \omega$) $|\alpha + \alpha + 1| = |\alpha + 1|$. However, $\lambda\alpha.\alpha^2$ is not continuous (this is exercise 107(ii)), and so the argument breaks down at limit ordinals. But this is retrievable.

Consider the following (origami proof) ordering. Order $\{\beta : \beta < \alpha\} \times \{\beta : \beta < \alpha\}$ as follows. Order pairs in the graph of $>$ lexicographically, so that if $\beta > \gamma$ and $\beta' > \gamma$, then put $\langle\beta,\gamma\rangle$ earlier than $\langle\beta',\gamma'\rangle$ iff $\beta < \beta'$ or $\beta = \beta' \wedge \gamma < \gamma'$. (That is what the vertical lines in the bottom right half of figure 8.1 are doing.)

Order pairs in the graph of $\leq$ in the colex ordering, so that if $\beta \leq \gamma$, and $\beta' \leq \gamma$, then put $\langle\beta,\gamma\rangle$ earlier than $\langle\beta',\gamma'\rangle$ iff $\gamma < \gamma'$ or $\gamma = \gamma' \wedge \beta < \beta'$. (That is what the horizontal lines in the top left half of figure 8.1 are doing.) At this point we have two disjoint sets, both well-ordered, and the operation of flipping ordered pairs is an isomorphism between them.

Then place every pair next to its flip. (That is to say, fold the top left corner down onto the bottom right corner.)

This is almost a well-ordering, but we have $\langle\beta,\gamma\rangle$ and $\langle\gamma,\beta\rangle$ sitting on top of each other, so it is not antisymmetric. Ordain that in each case the first of these two pairs shall be the one that is in the graph of $<$. If we interleave two well-orderings of length λ for λ limit, we clearly get a well-ordering of length λ as a result. If we interleave two well-orderings of length $\lambda + n$, we get a well-ordering of length $\lambda + 2n$.

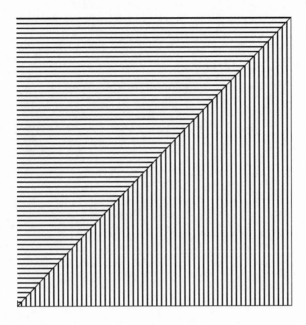

Fig. 8.1. $|\alpha^2| = |\alpha|$

Let us say that this well-ordering of $\{\beta : \beta < \alpha\} \times \{\beta : \beta < \alpha\}$ is of length $f(\alpha)$. f is a *continuous* function $f : On \to On$ such that $f(\alpha+1) = f(\alpha)+\alpha\cdot2+1$. This will have the property that $|f(\alpha)| = |\alpha^2|$. Because f is continuous, we will be able to prove by induction on α that $(\forall\alpha)(|\alpha| = |f(\alpha)|)$. But – because we know that $|f(\alpha)| = |\alpha^2|$ – the proof is now complete. ∎

LEMMA 91 *Bernstein's lemma.*

In figure 8.2 we see a representation of a set of size $x \cdot y$ split into two pieces of sizes a and b. Consider the U-shaped area labelled 'b', and its projection onto the horizontal axis. Does it cover the whole of the horizontal axis? If it does, then $b \geq^* y$. If it does not, then there is a line parallel to the other axis lying entirely within the complement of b, namely, a, whence $x \leq a$.[1] So we have proved

---

[1] Students have told me that this construction reminds them of a game called *Block-busters*.

$$(x \cdot y = a + b) \rightarrow (b \geq^* y \vee x \leq a),$$

which is Bernstein's Lemma.   ∎

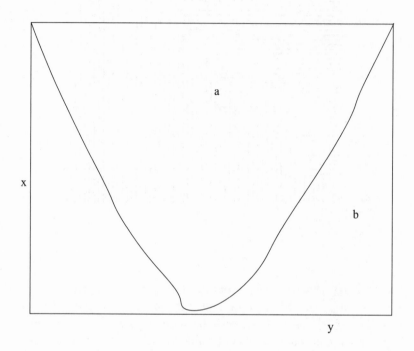

Fig. 8.2. Bernstein's lemma

EXERCISE 128

(i) *The proof in section 2.1.6.2 that there are $2^{\aleph_0}$ reals has a wrinkle caused by the fact that some rationals have more than one binary representation. Use Bernstein's lemma to iron it out.*

(ii) *There is a version of Bernstein's lemma that involves chopping into two pieces not a set of size $x \cdot y$ but rather a set that is the union of x pieces each of size at most y. (It does not have such an appealing picture associated with it.) State it correctly.*

DEFINITION 92

    (i) *An* **aleph** *is a cardinal of a well-ordered set.* $\aleph(\alpha)$ *is the least aleph* $\not\leq \alpha$.

    (ii) *We write '*$|A| \leq^* |B|$*' when there is a surjection of B onto A (or A is empty).*

    (iii) $\aleph^*(\alpha)$ *is the least aleph that is not* $\leq^* \alpha$.

    (iv) *If* κ *is an aleph, then* κ^+ *is the next aleph, which is of course the same as* $\aleph(\kappa)$.

There is no notation for the first ordinal that is not the length of a well-ordering of any set of size $\leq \alpha$. If we want '$\aleph(\alpha)$' to denote this object, we really do have to exploit the nasty hacky identification of cardinals with initial ordinals.

We had better show that $\aleph(\alpha)$ is always defined! The collection of alephs is naturally well-ordered, since to each aleph there corresponds an initial ordinal, so there is no problem about leastness. Existence needs to be checked. We do that next.

THEOREM 93 *(Hartogs-Sierpinski)*

    (i) $\aleph(\alpha)$ *is always defined (Hartogs);*

    (ii) $\aleph(\alpha) < 2^{2^{\alpha^2}}$ *(Sierpinksi).*

Let A be a set of size α. Every well-ordering of any subset of A is an element of $\mathcal{P}(A \times A)$. Therefore, the set of all well-orderings of subsets of A, is a subset of $\mathcal{P}(A \times A)$. Therefore the quotient, the set of all isomorphism classes of well-orderings of subsets of A, is a quotient of $\mathcal{P}(A \times A)$ and accordingly injects into $\mathcal{P}^2(A \times A)$. This structure is naturally well-ordered to the length of the sup of the lengths of the well-orderings represented in it, namely, the smallest ordinal not the length of any well-ordering of A. ∎

The axioms of ZF have the pleasing feature that, if we ever succeed in proving the existence of set with a given property, then by keeping track of the axioms we have used in the proof we can usually read off an upper bound for the size of the set whose existence we have proved. Use of the power set axiom contributes an exponentiation to the expression for the boud (which is what has just happened here) and use of the axiom of replacement contributes a 'sup'.

EXERCISE 129 *By coding a well-ordering as the set of its initial segments, show how to prove the following variant of Hartogs-Sierpinski:*

$$\aleph(\alpha) < 2^{2^{2^{\alpha}}}.$$

Recalling the definition of the the cardinal-valued function $\aleph^*(\alpha)$ from definition 92, find some upper bounds and some $\leq^*$-upper bounds for $\aleph(\alpha)$ and $\aleph^*(\alpha)$.

One immediate and standard application of Hartogs' is the following:

REMARK 94 *The axiom of choice is equivalent to the assertion that for all cardinals α and β, $\alpha \leq \beta \vee \beta \leq \alpha$.*

Proof: We established in exercise 26 that Zorn's lemma implied that for any two sets A and B there is an injection from one into the other. Zorn's lemma is an equivalent of the axiom of choice, so this gives us the left-to-right implication. For the other direction use Hartogs' theorem: compare α and $\aleph(\alpha)$. We must have $\alpha \leq \aleph(\alpha)$, so α must be an aleph. ∎

The continuum hypothesis (CH) is the assertion that there is no cardinal strictly between $\aleph_0$ and $2^{\aleph_0}$. GCH is the generalised continuum hypothesis – the assertion that for **all** cardinals α (not just $\aleph_0$) there are no cardinals strictly between α and 2^{α}.

EXERCISE 130

  (i) *Deduce AC from GCH;*
  (ii) *It is not provable without AC that every infinite set has a countable subset, but every infinite subset of $\Re$ has a countable partition.*

This corollary is not significantly harder than the preceding exercise, but a worked proof of a standard result may be helpful.

COROLLARY 95 *If $\alpha = \alpha^2$ for all cardinals α, then AC.*

Proof: Assume the hypothesis, so $(\alpha = \alpha^2)$ for all cardinals α. Now let α be a cardinal that is not an aleph. Then we have

$$(\alpha + \aleph(\alpha))^2 = (\alpha + \aleph(\alpha)).$$

Expand the left-hand side to get

$$\alpha^2 + 2 \cdot \alpha \cdot \aleph(\alpha) + (\aleph(\alpha))^2.$$

which can be simplified progessively to

$$\alpha + 2 \cdot \alpha \cdot \aleph(\alpha) + \aleph(\alpha)$$

(using ($\forall$ cardinals α)($\alpha = \alpha^2$)) and then (since if $\alpha = \alpha^2$, then certainly $\alpha = 2 \cdot \alpha$)

$$\alpha + \alpha \cdot \aleph(\alpha) + \aleph(\alpha),$$

which eventually becomes

$$\alpha \cdot \aleph(\alpha).$$

Then

$$\alpha \cdot \aleph(\alpha) = \alpha + \aleph(\alpha),$$

and we can use Bernstein's lemma to infer $\alpha \leq^* \aleph(\alpha) \vee \aleph(\alpha) \leq \alpha$. The second disjunct cannot happen (by definition of $\aleph(\alpha)$) and the first implies that α is an aleph.   ∎

COROLLARY 96 *AC is equivalent to the assertion that $\alpha = \alpha^2$ for all infinite cardinals.*

In fact, AC follows even from the apparently much weaker assertion that squaring of cardinals is merely injective. But that – like so many theorems about what AC is equivalent to – is beyond the scope of this book. However, we will have to consider what happens once we adopt AC as an axiom.

8.7.1 Cardinal arithmetic with the axiom of choice

Cardinal arithmetic with the axiom of choice is very simple. Although there are still some questions about exponentiation, all questions about $\cdot$ and $+$ are easily answered by the following result.

THEOREM 97 *(AC) For all cardinals α and β, $\alpha + \beta = \alpha \cdot \beta = max(\alpha, \beta)$.*

Proof: The axiom of choice implies that every set can be well-ordered, so every cardinal is an aleph. All well-orderings are comparable in length by proposition 62; so all alephs are comparable in size; so AC implies that $\leq$ on cardinals is a total order.

Now, given two cardinals α and β, one of them must be larger – β, say. Then we have $\beta \leq \beta + \alpha \leq \beta + \beta \leq \beta \cdot \beta = \beta$.

Similarly, with $\alpha \leq \beta$, we have $\beta \leq \beta \cdot \alpha \leq \beta \cdot \beta = \beta$. ∎

We have just sewn up the theory of $\cdot$ and $+$. In contrast, there is very little we can prove about exponentiation. We have just seen CH and GCH. It is now well-known that these propositions cannot be established or refuted from uncontroversial axioms like ZF + AC. To see what can be done, let us return to Bernstein's lemma. A simple induction on $\mathbb{N}$ shows us that, if $\{A_i : i < n\}$ and $\{B_i : i < n\}$ are families of sets with a map from the union of the B's onto the product of the A's then something happens. With AC we can even get a result on infinite sums and products.

THEOREM 98 *(The Jordan-König theorem (AC))* If $\{A_i : i \in I\}$ and $\{B_i : i \in I\}$ *are families of sets such that* $(\forall i \in I)(|A_i| \not\geq^* |B_i|)$, *then*

$$|\bigcup_{i\in I} A_i| \not\geq^* |\prod_{i\in I} B_i|.$$

Proof: Suppose not, and that $f : \bigcup_{i\in I} A_i \to \prod_{i\in I} B_i$. We show that f is not onto. For each $i \in I$, let $f_i : A_i \to B_i$ be $\lambda x_{A_i}.(f(x)(i))$. f_i cannot be onto by hypothesis, so pick (remember we are using AC) n_i to be a member of $B_i \setminus f_i``A_i$. Now we find that the function $\lambda i.n_i$ is not in the range of f, for otherwise, if $f(a) = \lambda i.n_i$, where $a \in A_i$ say, then $f_i(a) = (\lambda x.(f(x))(i))(a) = (f(a))(i) = (\lambda i.n_i)(i) = n_i$, contradicting our choice of n_i. ∎

The Jordan-König theorem is equivalent to AC, because it implies that the product of nonempty sets is nonempty.

COROLLARY 99 $2^{\aleph_0} \neq \aleph_\omega$.

Proof: Take the A_i to be of size $\aleph_i$, for $i \in \mathbb{N}$, and the B_i to be all of size $2^{\aleph_0}$ and note that $(2^{\aleph_0})^{\aleph_0} = 2^{\aleph_0}$. ∎

EXERCISE 131 *Prove that* $\alpha < \alpha^{cf(\alpha)}$.

(Writing '$\alpha < \alpha^{cf(\alpha)}$' like this is a bit slovenly. Clearly, α and $\alpha^{cf(\alpha)}$ are to be cardinals, so the exponent, $cf(\alpha)$, has to be a cardinal too. But it is *ordinals* that have cofinalities, not cardinals, and cofinalities are ordinals and not cardinals! We could have exploited the overloading of vertical bars to write '$|cf(\alpha)|$', but this is not idiomatic. It is usual to write what I wrote above, thereby exploiting the tacit identification of the cardinal α with the initial ordinal corresponding to it, and the similar identification of the ordinal $cf(\alpha)$ with the corresponding aleph. Slovenly it may be, but it is universally practiced.)

The Jordan-König theorem is close to the limit of what one can prove about exponentiation by elementary methods. With the advent of Cohen's method of *forcing* in the 1960s, Easton (1970) was able to show that the function that accepted an ordinal α and returned that β such that $2^{\aleph_\alpha} = \aleph_\beta$ could be (roughly) anything – as long as α was regular, but it told us nothing about what $2^{\aleph_\alpha}$ could be if α was singular. Silver (1974) proved a theorem about this in 1974, which does not use forcing, and could in principle be comprehended by a reader of a text at this level, but it is hard and is best motivated in the context of Easton's work – which cannot be treated here.

8.8 Independence proofs

Although clearly some instances of the axiom schemes of separation and replacement can be derived from others, it is standard that the remaining axioms of ZF are independent from each other. For any other axiom A we can show that $ZF \setminus A \not\vdash A$. And for replacement we can show that $ZF\setminus$ replacement does not imply all instances of replacement, though it does prove some.

A device that turns up in many of these independence proofs is the idea of the set of things that are hereditarily ϕ, where ϕ is a one-place predicate. The intuition is that x is hereditarily ϕ if everything in $TC(x)$ is ϕ. If you are ϕ and all your members are hereditarily ϕ, then you are hereditarily ϕ, whence the rectype definitions:

DEFINITION 100

$$\mathcal{P}_\kappa(x) := \{y \subseteq x : |y| < \kappa\}; \ H_\kappa := \bigcap\{y : \mathcal{P}_\kappa(y) \subseteq y\};$$
$$\mathcal{P}_\phi(x) := \{y \subseteq x : \phi(y)\}; \ \ H_\phi := \bigcap\{y : \mathcal{P}_\phi(y) \subseteq y\}.$$

A word is in order on the definition and the notation involved. The use of the set-forming bracket inside the '$\bigcap$' is naughty: in general there is no reason to suppose that the collection of all y such that $\mathcal{P}_\phi(y) \subseteq y$ is a set. However, its intersection will be a set – as long as it is nonempty! And if there is even one x such that $\mathcal{P}_\phi(x) \subseteq x$, then $\{y \subseteq x : \mathcal{P}_\phi(y) \subseteq y\}$ will have the same intersection as $\{y : \mathcal{P}_\phi(y) \subseteq y\}$, and so no harm is done. But this depends on there being such an x. If there is, we are in the same situation we were in with the implementation of $\mathbb{N}$. If not, then the collection H_ϕ will be a proper class, and we have to define it as the collection of those x with the property that everything in $TC(x)$ is of size $< \kappa$. If H_ϕ is a set, then the two definitions are of course equivalent; but if it is not, it is only the definition in terms of TC that works. The definition in terms of TC is the standard one, but I find that my definition is more helpful to people who are used to thinking in terms of inductive definitions. After all, H_ϕ is a rectype. It has an empty set of founders and one (infinitary!) constructor that says that a subset of H_ϕ that is itself ϕ is also in H_ϕ.

WF is just $H_{x=x}$.

EXERCISE 132

(i) *Show that, if κ is regular and we have AC, then we can take H_κ to be the set of x s.t. $|TC(x)| < \kappa$.*

(ii) *Show that the collection of hereditarily well-ordered sets is not a set.*

REMARK 101 *If $\phi(x) \to \phi(f\,{}^{\shortmid\shortmid}x)$ for all x and f, then H_ϕ is a model for replacement.*

Proof: For H_ϕ to be a model of replacement it is sufficient that, if $x \in H_\phi$ and $f : H_\phi \to H_\phi$ is defined by a formula with parameters from H_ϕ only, all of whose quantifiers are restricted to H_ϕ, then $f\,{}^{\shortmid\shortmid}x$ is also in H_ϕ. But this condition is met because by assumption a surjective image of a set that is ϕ is also ϕ: indeed, we did not even need the italicised condition. ∎

8.8.1 Replacement

The only independence proof that we will give which does not use an H_ϕ is the independence of the axiom scheme of replacement. Let us get it out of the way now.

$V_{\omega+\omega}$ is a model for all the axioms except replacement. It contains well-orderings of length ω but cannot contain $\{V_{\omega+n} : n \in \mathbb{N}\}$ because we can use the axiom of sumset (and $V_{\omega+\omega}$ is clearly a model for the axiom of sumset!) to get $V_{\omega+\omega}$.

Readers are encouraged to check the details for themselves to gain familiarity with the techniques involved.

8.8.2 Power set

H_κ is never a model for power set unless κ is strong limit. Although it is not obvious, H_κ is always a set according to ZF. The proof does not need (much) AC, but the axiom of foundation is essential.[1]

THEOREM 102 *If κ is an aleph, $|H_{\kappa^+}| \le 2^\kappa$.*

Proof: Let X be a set of size 2^κ. Assume enough AC to be sure that $|X| = |\mathcal{P}_{\kappa^+}(X)|$ because κ^+ is well-behaved. We need a bit of choice to do this because $\kappa^2 = \kappa$ is not enough. For example, there are $2^{\aleph_0}$ ω-sequences of reals, since there are $(2^{\aleph_0})^{\aleph_0} = 2^{\aleph_0 \cdot \aleph_0} = 2^{\aleph_0}$, but that does not tell us there are precisely $2^{\aleph_0}$ countable sets of reals. There are clearly $\ge 2^{\aleph_0}$ countable sets of reals, and the argument we have just sketched shows that there are $\le^* 2^{\aleph_0}$ countable sets of reals. To infer that there are precisely $2^{\aleph_0}$ countable sets of reals from the fact that there are $2^{\aleph_0}$ ω-sequences of reals we would need to be able to pick, for each countable set of reals, a well-ordering of it to length ω.

Let us fix an injection $\pi : \mathcal{P}_{\kappa^+}(X) \hookrightarrow X$. We construct an injection $h : H_{\kappa^+} \hookrightarrow X$ by recursion thus: $h(x) := \pi(h``x)$. By considering a member of H_{κ^+} of minimal rank not in the range of h, we show easily that h is total. It is injective because π is one – one. The range of h is a set by comprehension, and so its domain (which is H_{κ^+}) is also a set, by replacement. ∎

REMARK 103 $|H_{\aleph_1}| = 2^{\aleph_0}$

Proof: We have just seen $|H_{\aleph_1}| \le 2^{\aleph_0}$. The other direction follows immediately from the fact that $V_{\omega+1}$ is a subset of $H_{\aleph_1}$ of size $2^{\aleph_0}$. ∎

There is another way of proving that $H_{\aleph_1}$ is a set. Recall that $\lambda x.\mathcal{P}_{\aleph_1}(x)$ is not ω-continuous. If you think about this for a while, you will realise that this function is α-continuous for those α such that $cf(\alpha) > \omega$. The first such ordinal is ω_1. (Look back at remark 73.) So all we have to do is iterate this function ω_1 times and we will reach a fixed point. $H_{\aleph_1}$ will be a subset of this fixed point and will be a set by comprehension.

---

[1] We will see soon that, if we do not assume the axiom of foundation, we can easily construct models containing as many Quine atoms (sets $x = \{x\}$) as we want. Since these objects are clearly hereditarily of size less than κ^+, there is no point in asking about the size or sethood of $H_{\aleph_1}$ unless we assume some form of foundation.

$H_{\aleph_1}$ gives us a model of ZF minus the power set axiom. The axiom of infinity will hold because there are genuinely infinite sets in $H_{\aleph_1}$. This is not sufficient by itself as "is infinite" is not Δ_0, but whenever X is such a set there will be a bijection from X onto a proper subset of itself, and this bijection (at least if our ordered pairs are Wiener-Kuratowski) will be a hereditarily countable set. So any actually infinite member of $H_{\aleph_1}$ will be believed by $H_{\aleph_1}$ to be actually infinite. We have been assuming the axiom of choice, so the union of countable many elements of $H_{\aleph_1}$ is also an element of $H_{\aleph_1}$, so it is a model of the axiom of sumset.

Everything in $H_{\aleph_1}$ is countable and therefore well-ordered, and, under most implementations of pairing functions, the well-orderings will be in $H_{\aleph_1}$, too, so $H_{\aleph_1}$ is a model of AC, even if AC was not true in the model in which we start.

8.8.3 Independence of the axiom of infinity

$H_{\aleph_0}$ provides a model for all the axioms of ZF except infinity and thereby proves the independence of the axiom of infinity. (We constructed a copy of $H_{\aleph_0}$ on page 40.)

The status of AC in $H_{\aleph_0}$ is like its status in $H_{\aleph_1}$. Everything in $H_{\aleph_0}$ is finite and therefore well-ordered, and under most implementations of pairing functions the well-orderings will be in $H_{\aleph_0}$ too, so $H_{\aleph_0}$ is a model of AC, even if AC was not true in the model in which we start. This is in contrast to the situation obtaining with the countermodels to sumset and foundation: the truth-value of AC in those models is the same as its truth-value in the model in which we start.

8.8.4 Sumset

$\beth$ ("Beth") numbers are defined by setting $\beth_\alpha := |V_\alpha|$ or, recursively, by $\beth_0 := \aleph_0; \beth_{\alpha+1} := 2^{\beth_\alpha}$, taking sups at limits. Let us for the moment say that a set of size less that $\beth_\omega$ is **small**.

Then $H_{\beth_\omega}$, the collection of hereditarily small sets, proves the independence of the axiom of sumset. This is because there are well-orderings of length $\omega + \omega$ inside $V_{\omega+n}$ for n small, so by replacement $\{V_\alpha : \alpha < \omega + \omega\}$ is a set. Indeed it is a hereditarily small set. But $\bigcup \{V_\alpha : \alpha < \omega + \omega\}$ is not hereditarily small, being of size $\beth_\omega$.

EXERCISE 133 *Establish that the collection of hereditarily small sets is a set.*

8.8.5 Foundation

For the independence of the axiom of foundation and the axiom of choice we need a Rieger-Bernays model for independence of foundation.

If $\langle V, R \rangle$ is a structure for the language of set theory, and π is any permutation of V, then we say $x \ R_\pi \ y$ iff $x \ R \ \pi(y)$. $\langle V, R_\pi \rangle$ is a *permutation model* of $\langle V, R \rangle$. We call it V^π. Alternatively, we could define Φ^π as the result of replacing every atomic wff $x \in y$ in Φ by $x \in \pi(y)$. We do not rewrite equations in this operation: $=$ is a logical constant, not a predicate letter. The result of our definitions is that $\langle V, R \rangle \models \Phi^\pi$ iff $\langle V, R_\pi \rangle \models \Phi$. Although it is possible to give a more general treatment, we will keep things simple by using only permutations whose graphs are sets.

A wff ϕ is stratified iff we can find a *stratification* for it, namely, a map f from its variables (after relettering where appropriate) to $\mathbb{N}$ such that, if the atomic wff '$x = y$' occurs in ϕ, then $f('x') = f('y')$, and if '$x \in y$' occurs in ϕ, then $f('y') = f('x') + 1$.

To discuss these topics properly we will also need the notation $j =_{\mathrm{df}}$ $\lambda f \lambda x.(f``x)$. The map j is a group homomorphism: $j(\pi\sigma) = (j\pi)(j\sigma)$.

We shall start with a lemma and a definition, both due to Henson (1973). The definition arises from the need to tidy up Φ^τ. A given occurrence of a variable 'x' that occurs in 'Φ^τ' may be prefixed by 'τ' or not, depending on whether or not that particular occurrence of 'x' is after an '$\in$'. This is messy. If there were a family of rewriting rules around that we could use to replace $x \in \tau(y)$ by $\sigma(x) \in \gamma(y)$ for various other σ and γ, then we might be able to rewrite our atomic subformulæ to such an extent that, for each variable, all its occurences have the same prefix.

Why bother? Because once a formula has been coerced into this form, every time we find a quantifier Qy in it, we know that all occurrences of y within its scope have the same prefix. As long as that prefix denotes a permutation, we can simply remove the prefixes! This is because $(Qx)(\ldots \sigma(x) \ldots)$ is the same as $(Qx)(\ldots x \ldots)$. If we can do this for all variables, then τ has disappeared completely from our calculations and we have an invariance result. When can we do this?

Henson's insight was as follows. Suppose we have a stratification for Φ and permutations τ_n (for all n used in the stratification) related somehow to τ, so that, for each n,

$$x \in \tau(y) \longleftrightarrow \tau_n(x) \in \tau_{n+1}(y).$$

Then, by replacing ' $x \in \tau(y)$ ' by ' $\tau_n(x) \in \tau_{n+1}(y)$ ' whenever 'x' has been assigned the subscript n, every occurrence of 'x' in ' Φ^τ ' will have the same prefix. Next we will want to know that τ_n is a permutation, so that in any wff in which 'x' occurs bound – $(\forall x)(\ldots \tau_n(x) \ldots)$ – it can be relettered $(\forall x)(\ldots x \ldots)$ so that 'τ' has been eliminated from the bound variables. It is not hard to check that the definition that we need to make this work is as follows:

DEFINITION 104 $\tau_0 = $ identity; $\tau_{n+1} = (j^n \tau) \circ \tau_n$.

This definition is satisfactory as long as $j^n(\tau)$ is always a permutation of V whenever τ is, for each n. But if the graph of τ is a set, we need have no worries on that score. This gives us immediately a proof of the following result.

LEMMA 105 *(Henson (1973)) Let Φ be stratified with free variables 'x_1', ..., 'x_n', where 'x_i' has been assigned an integer k_i in some stratification. Let τ be a setlike permutation and V any model of NF. Then*

$$(\forall \vec{x})V \models (\Phi(\vec{x})^\tau \longleftrightarrow \Phi(\tau_{k_1}(x_1) \ldots \tau_{k_n}(x_n))).$$

In the case where Φ is closed and stratified, we infer that if τ is a permutation whose graph is a set, then

$$V \models \Phi \longleftrightarrow \Phi^\tau.$$

REMARK 106 *(Scott (1962)) If $\langle V, \in \rangle \models$ ZF and τ^{-1} is a permutation of V whose graph is a set, then $\langle V, \in_\tau \rangle \models$ ZF.*

Proof: The stratified axioms are no problem. The only unstratified axiom that we wish to make true in the model is the axiom scheme of replacement. It is easy enough to check for any ϕ that, if $\forall x \exists ! y \phi$, then $\forall x \exists ! y \phi^\tau$, so that for any set X the image of X in ϕ^τ is also a set. Call it Y. But then $\tau^{-1}(Y)$ is the image-of-X-under-ϕ (in the sense of V^τ). ∎

We now take π to be the transposition $(\emptyset, \{\emptyset\})$. In $\mathfrak{M}^\pi$ the old empty set has become a Quine atom: an object identical to its own singleton: $x \in_\pi \emptyset \longleftrightarrow x \in \pi(\emptyset) = \{\emptyset\}$. So $x \in_\pi \emptyset \longleftrightarrow x = \emptyset$. So $\mathfrak{M}^\pi$ is a model for all the axioms of ZF except foundation.

8.8.5.1 Antifoundation

There is another proof of the independence of the axiom of foundation, which goes back to work of Forti and Honsell (1983). Under this approach a set is regarded as an isomorphism class of accessible pointed digraphs (APGs). An APG is a digraph with a designated vertex v such that every vertex has a dipath reaching v.

The best-known exposition of this material is the eminently readable Aczel (1988). I shall not treat it further here, since – although attractive – it is recondite, and the proof of independence of foundation that it gives does not (unlike the previous one) naturally give rise to a proof of the independence of the axiom of choice. This is our next chore.

8.8.6 Choice

We start with a model of ZF + foundation and use Rieger-Bernays model methods to obtain a permutation model with a countable set A of Quine atoms. The permutation we use to achieve this is the product of all transpositions $(n, \{n\})$ for $n \in \mathbb{N}^+$. A will be a **basis** for the ill-founded sets in the sense that any class X lacking an $\in$-minimal element contains a member of A. Since the elements of A are Quine atoms, every permutation of A is an $\in$-automorphism of A, and since they form a basis we can extend any permutation σ of A to a unique $\in$-automorphism of V in the obvious way: set $\sigma(x) := \sigma``x$. Notice that the collection of sets that this definition does not reach has no $\in$-minimal member if nonempty, and so it must contain a Quine atom. But σ by hypothesis is defined on Quine atoms. (a, b) is of course the transposition swapping a and b, and we will also write it for the unique automorphism to which the transposition (a, b) extends.

Every set x gives rise to an equivalence relation on atoms. Say $a \sim_x b$ if (a, b) fixes x. We say x is of (or has) **finite support** if $\sim_x$ has a cofinite equivalence class. (If it has a cofinite equivalence class, it can have only one, and those remaining will all be finite.) The union of the (finitely many) remaining (finite) equivalence classes is the **support** of x. Does that mean that x is of finite support iff the transitive closure $TC(x)$ contains finitely many atoms? Well, if $TC(x)$ contains only finitely many atoms, then x is of finite support (x clearly cannot tell apart the cofinitely many atoms not in $TC(x)$), but the converse is not true: x can be of finite support if $TC(x)$ contains cofinitely many atoms. (Though that is not a sufficient condition for x to be of finite support!!)[1]

It would be nice if the class of sets of finite support gave us a model of something sensible but extensionality fails: if A is of finite support, then $\mathcal{P}(A)$ and the set $\{X \subseteq A : X \text{ is of finite support}\}$ are both of finite support and have the same members with finite support. We have to consider the class of elements hereditarily of finite support. Let us call it HF. This time we do get a model of ZF.

Let f be a definable function. Notice that, if (a, b) fixes every argument to f, it must also fix its values, by single-valuedness of f. This has the immediate consequence that HF is closed under all definable operations: sets that are of finite support are of finite support in virtue of a cofinite set of atoms that they cannot discriminate. So, if $x_1 \ldots x_n$ are all of finite support, then $f(x_1 \ldots x_n)$ is in HF in virtue of the intersection of the cofinite sets of atoms associated with $x_1 \ldots x_n$, and the intersection of finitely many cofinite sets is cofinite. This takes care of all the axioms of ZF except infinity. Since every well-founded set is fixed under all automorphisms, HF will contain all well-founded sets, so since there was an infinite well-founded set in the model we started with, HF will contain that infinite set and will model infinity. Finally, HF satisfies replacement because of remark 101.

We now have a very simple independence proof of AC from ZF. Consider the set of (unordered) pairs of atoms. This set is in HF. But clearly no selection function for it can be in HF. Suppose f is a selection function. It picks a (say) from $\{a, b\}$. Then f is not fixed by (a, b). Clearly the equivalence classes of $\sim_f$ are going to be singletons, and $\sim_f$ is going to be of infinite index and f is not of finite support.

So the axiom of choice for countable sets of pairs fails. Since this axiom is about the weakest version of AC known to man, this is pretty good. The slight drawback is that we have had to drop foundation to achieve it. On the other hand, the failure of foundation is not terribly grave: the only ill-founded sets are those with a Quine atom in their transitive closure s, so there are no sets that are gratuitously ill-founded: there is a basis of countably many Quine atoms.

---

[1] A counterexample: well-order cofinitely many atoms. The graph of the well-order has cofinitely many atoms in its transitive closure, but they are all inequivalent.

8.9 The axiom of choice

In Russell's *Introduction to Mathematical Philosophy* (p.126) is the wonderful parable of the millionaire whose wardrobe contains a countable infinity of pairs of shoes and a countable infinity of pairs of socks. It is usually felt to be obvious that there are countably many shoes and countably many socks in his attic: that is to say, it is obvious that there is a bijection between the shoes and the natural numbers, and a bijection between the socks and the natural numbers. As the great logician Dana Scott said (1962), "Nothing supports belief like proof", and with a bit of prodding most students can be persuaded to say that the left shoe from the nth pair can be sent to $2n$ and the right shoe from the nth pair can be sent to $2n + 1$. This indeed shows that there are countably many shoes. "And the socks?" one asks. With luck the student will reply that the same technique will work, whereupon they can be ribbed for wearing odd socks. Indeed, for this strategy to work, it is not only necessary that all (but finitely many) of the pairs of socks be odd, but that they be odd in some uniform way, such as one red and one black. The fact that there is – or ought to be – no systematic way of choosing one sock from each pair means that we need an axiom saying that there is a choice function – even if we cannot give one explicitly.

One is tempted to say, "The example of the socks proves that it's obvious that the union of countably many pairs is countable, and if we need the axiom of choice to prove it, then the axiom of choice we'd better have". The point is not that this is a fallacy of affirming the consequent: there is nothing wrong with arguing for an axiom on the grounds that it has a lot of obviously true consequences that do not appear to follow from the other axioms we have settled on. We saw this in connection with lemma 85 and the axiom scheme of replacement. The difference here is that this benign use of affirming the consequent does not help: the propositions that are obviously true and appear to follow from the axiom of choice are not in fact nontrivial consequences of the axiom of choice at all and cannot be used to argue for it.

An illustration may help. Any subset of the plane or of $\Re^3$ that is a union of countably many pairs is indeed countable, but that is not the same as saying that any union of countably many pairs is countable. If the socks from countably many pairs of socks are dispersed through $\Re^3$, then the interior of every sock must contain a rational number, and there will be a least rational number inside each sock, and this can be used to count the socks. So this is a proof that there are countably many socks in countably many pairs of socks:[1] it is not a proof of the axiom of choice for countably many pairs.

Another – common and important – example of a spuriously plausible assertion is the claim that a union of countably many countable sets is countable. The most illuminating discussion of this known to me is one I learned from Conway (oral tradition). Conway distinguishes between a **counted set**, which is a structure $\langle X, f \rangle$ consisting of a set X with a bijection f onto $\mathbb{N}$, and a **countable set**, which is a naked set that just happens to be the same size as $\mathbb{N}$. As Conway says, elliptically but memorably: a counted union[2] of counted sets is counted; a countable union of counted sets is countable, but a counted union of countable sets, and *a fortiori* a countable union of countable sets could – on the face of it – be anything under the sun. The fact that is obvious is not 'a countable union of countable sets is countable' but the quite distinct 'a counted union of counted sets is counted'.

Put like this, it sounds as if failures of the axiom of choice happen only when we have imperfect information about sets. If we were God, we would be able to well-order the universe and we would be able to see that a union of countably many countable sets is countable. What else could it possibly be? The counterexample we have contrived seems indeed contrived, and to happen only because we cannot tell Quine atoms apart. But God can, so God knows that AC is true. In philosophical terminology, people who believe in the existence of entities of flavour $\mathcal{F}$ are **realists about** $\mathcal{F}$s. Realists about mathematical entities are often called **platonists**. It certainly seems to be the case that platonists tend to believe the axiom of choice. They believe it for substantially the same reasons that God knows that AC is true. If mathematical objects are real, then questions about their sizes must have real answers. The only possible answer to the question about the number of socks seems to be $\aleph_0$, and if the only way to infer that is to assume AC, then realists have a good reason to believe AC.

8.9.1 AC and constructive reasoning

The idea that independence of AC is connected with incomplete information should remind us of constructive reasoning: when we reason

---

[1] Or, more correctly, we can prove the following without any use of AC at all: if A is a family of disjoint subsets of $\Re$, each with nonempty interior, and A has a countably infinite partition into pairs, then A is countable.

[2] We overload 'union' here.

constructively we deliberately refrain from exploiting certain kinds of information. This would lead us to expect that AC should sit ill with constructive reasoning. This turns out to be the case: the axiom of choice implies the law of excluded middle. Indeed, the law of excluded middle follows if there is even one nontrivial well-founded relation.

8.9.2 The consistency of the axiom of choice?

The idea that, if we have perfect information about sets, we can well-order them gives rise to an idea for a consistency proof for the axiom of choice. Recall the rectype WF: its sole constructor adds at each stage *arbitrary* sets of what has been constructed at earlier stages. If we modify the construction so that at each stage we add only those sets-of-what-has-been-constructed-so-far about which we have a great deal of information, then with luck we will end up with a model in which every set has a description of some sort, and in which we can distinguish socks *ad lib.*, and in which the axiom of choice is true. This strategy can be made to work, but there is no space for all the details here.

Other approaches

Readers should not form the impression that the "what do we trust?" approach that leads to the cumulative hierarchy is the only sensible response to the paradoxes: we might decide to trust syntax rather than creation. What does this mean? For most readers the most striking feature of the Russell class is not the fact that the cumulative hierarchy does not construct it, but rather the dangerous paradoxical reasoning involved and the obvious parallels with the other paradoxes. Perhaps there are other solutions to the paradoxes to be found by following up this syntactic insight? One might feel that sets ought to be naturally regimented into layers in such a way that the questions like $x \in x$? are somehow illegitimate. One way of doing this would be to restrict the naïve set existence scheme

$$(\forall \vec{x})(\exists y)(\forall z)(z \in y \longleftrightarrow \Phi)$$

to those where Φ is *stratified* in the sense of section 8.8.5. It is now known that this approach successfully skirts the paradoxes – at least if we weaken extensionality to the extent of allowing distinct empty sets. (Two nonempty sets with the same members must still be the same set.) It is not known if this weakening of extensionality is needed. Why might anyone want to try this approach anyway? We noted on page 178 that the problem with paradoxical sets might not be their size but the descriptions they answer to. Stratified descriptions, with their flavour of avoiding self-reference, have a certain appeal. However, there is another reason for suspecting that this approach might have merit. We can think of Cantor's theorem as telling us something about models of set theory thought of in the di Giorgi way. In the di Giorgi view, the injection $i : A \hookrightarrow \mathcal{P}(A)$ can be thought of as a coding of subsets of A by elements of A. Cantor's theorem tells us that there are always subsets that are not coded. However, through i, one can make a choice of which subsets of A are to be coded and which not. Suppose we have a di Giorgi structure $\langle A, i \rangle$. If we now compose i on the right with a permutation π of A, we have a new di Giorgi structure, but it has made the same choices as the original structure did about which subsets of A are coded by elements of A. What properties of the model are invariant under this kind of monkeying around with permutations? The reader of section 8.8.5 will be prepared for the news that it is precisely the stratified properties that are preserved.

So stratification is at least a sensible mathematical notion. Whether a set theory based on restricting naïve comprehension to accomodate it is going to give rise to a fruitful alternative to ZF remains to be seen. The question has now been open for more than 60 years. Meanwhile, there are other syntactic tricks that one can try. It is known that, if one restricts naïve comprehension by not allowing negation signs inside Φ, then no contradiction can be proved. In fact, this constraint can be slightly relaxed, and the result is a set theory called GPC (generalised positive comprehension), but although this relaxation can be explained, the explanation is neither as elementary nor as satisfying as the story behind stratification.

I am not trying to persuade the reader that they should drop ZF and take up the study of stratified set theory or generalised positive comprehension. Even if you do not believe that set membership is a well-founded relation, the rectype WF remains an object worthy of study. I merely wish to make the point that alternatives to ZF are available and are legitimate objects of study too.

Exercises

EXERCISE 134 *Define E on* $\mathbb{N}$ *by: n E m iff the nth bit in the binary* expansion of m is 1. (Remember to start counting at the 0th bit!!) Do you recognise this structure?

EXERCISE 135 If you got that easily, consider the following more complicated version: n E_O m iff either m is even and the nth bit in the binary expansion of $m/2$ is 1 or m is odd and the nth bit in the binary expansion of $(m-1)/2$ is 0. You have almost certainly never seen this structure before: what can you say about it?

EXERCISE 136 An antimorphism is a permutation π of V so that $\forall x \; y \; x \in y \longleftrightarrow \pi(x) \notin \pi(y)$. Prove (without using the axiom of foundation) that no model of ZF has an antimorphism.

   (i) Find an antimorphism of the second structure in exercise 135.

   (ii) Is it unique? (Hint: Consider the dual of the preceding structure, i.e., the natural numbers with the relation n E_{O^*} m iff either m is even and the nth bit in the binary expansion of $m/2$ is 0 or m is odd and the nth bit in the binary expansion of $(m-1)/2$ is 1. Prove that this is isomorphic to the naturals with E_O.)

EXERCISE 137 Let X be a transitive set. If R is an equivalence relation on X and Y, Z are subsets of X, we can define $R'(Y, Z)$ iff $(\forall y \in Y)(\exists z \in Z)(R(y,z)) \wedge (\forall z \in Z)(\exists y \in Y)(R(y,z))$. Check that the restriction of R' to X is also an equivalence relation on X.

Show that this operation on equivalence relations has a fixed point. For any fixed point, one can take a quotient. Show how to define a membership relation on the quotient in a natural way and such that the result is a model of extensionality as well.

This construction is of particular interest if X is a V_α and the fixed point is the greatest fixed point. What can you say about the quotient in this case?

EXERCISE 138 Use AC to show that, if every chain in $\langle P, \leq_P \rangle$ has a sup, then every directed subset does too.

9

Answers to selected questions

Chapter 1: Definitions and notations

Exercise 1

There are 2^{n^2} binary relations on a set of size n, of which 2^{n^2-n} are reflexive, $2^{n \cdot (n-1)/2}$ are fuzzies, $2^{n \cdot (n+1)/2}$ are symmetrical, $2^n \cdot 3^{n \cdot (n-1)/2}$ are antisymmetric, $2^n \cdot 3^{n \cdot (n-1)/2}$ are trichotomous and $n!$ are total orders. The complement of a trichotomous relation is antisymmetric (and vice versa), which is why the numbers are the same.

To get a bijection between the set of antisymmetric trichotomous relations and the set of symmetrical relations, fix a total order of the underlying set. To obtain a matrix for an antisymmetric trichotomous relation from the matrix for an symmetrical relation, pointwise XOR the matrix with the matrix for the total order. There does not seem to be a natural bijection between the set of antisymmetric trichotomous relations and the set of symmetrical relations, merely a natural map from the set of total orders to the set of such bijections.

Exercise 3

Are the two following conditions on partial orders equivalent?

  (i) $(\forall xyz)(z < x \not\le y \not\le x \to z < y)$.
  (ii) $(\forall xyz)(z > x \not\le y \not\le x \to z > y)$.

They are. We will assume (i) and deduce (ii). To this end, assume $z > x$, $x \not\le y$ and $y \not\le x$ and hope to deduce $z > y$.

$x < z$ tells us that $z \not\le y$, for otherwise $x \le y$ by transitivity, contradicting the hypothesis. Next, assume the negation of what we are trying to prove. This gives us $y = z \lor y \not\le z$. The first disjunct is impossible

210

($x < z$ but $x \not\leq y$), but the second gives us $y \not\leq z \not\leq y$ and $x < z$, so by (i) we can infer $x < y$, contradicting the assumption.

The proof in the other direction is analogous.

Consider the strict partial order corresponding to a partial order satisfying this condition we have just been discussing. If it is well-founded, it is said to be a **pre-well-ordering**. This is because we can think of it as a strict total ordering of the equivalence classes (under the relation $x \simeq y$ iff $x = y \vee x \not\leq y \not\leq x$), and if $<$ is well-founded, this is in fact a well-ordering of the equivalence classes. If we drop the well-foundedness condition, we have a thing that should presumably be called a **pre-total-ordering**, but this terminology is not used.

Exercise 13

(i) is countable; (ii) is uncountable; (iii) is countable; (iv) is uncountable; (v) is uncountable; (vi) is countable; (vii) is uncountable; (viii) is countable.

Chapter 2: Recursive datatypes

Exercise 7

The parameter to watch is the ordered pair $\langle h, n \rangle$, where h is the highest common factor of the integers played so far by either player and n is the number of multiples of h that cannot be expressed as a sum-of-multiples of numbers played so far. Every move by either player transforms $\langle h, n \rangle$ to $\langle h', n' \rangle$, which is below $\langle h, n \rangle$ in the lexicographic order.

Have a look at http://www.monmouth.com/ colonel/sylver/

Exercise 8

Order $\mathbb{N}$ by putting all the odds in their usual order before all the evens in their usual order.

Exercise 10

Suppose *per impossibile* that there were a cube dissected into finitely many smaller cubes all of different sizes. Each face of the cube is now dissected into squares. Think about a square of smallest size on one of the faces. None of the cubes immediately behind it can overlap the edges, and so the faces of those cubes constitute a dissection of that square into smaller squares.

Exercise 11

First we show that $\trianglelefteq$ is well-founded.

If $X \subseteq \mathbb{N}$, then either there is $x \in X$ with $x > 100$, and any such x will be $\trianglelefteq$-minimal, or, if not, then the largest member of X is $\trianglelefteq$-minimal.

Now to show that f and g agree. Suppose $(\forall y)(y \trianglelefteq x \rightarrow f(y) = g(y))$. We wish to infer that $f(x) = g(x)$. If $x > 100$, this is true irrespective of the induction hypothesis, so we will go straight to the hard part, where $x \leq 100$. We want to prove that $f(f(x + 11)) = 91$.

If $x = 100$, then $f(x) := f(f(111))$. $f(111) = 101$ and $f(101) = 91$, which is what we wanted.

Notice that we have not used the induction hypothesis yet. It might be a good idea to start by trying to get a feeling for what the relation $\trianglelefteq$ does. It is the reverse of the usual ordering on numbers ≤ 100. We have got $f(100)$ sorted out, so let us orient ourselves by thinking about $f(99)$, since 99 is $\trianglelefteq$-minimal among the things we have not yet considered. This must be $f(f(110))$. $f(110)$ is 100, and we have just shown $f(100)$ to be 91.

So let us try the case $90 \leq x < 100$. Naturally, $f(x) := f(f(x + 11))$. But $x + 11 > 100$, so $f(x + 11) = (x + 11) - 10 = x + 1$. But $x + 1 \trianglelefteq x$ and so, by the induction hypothesis, $f(x + 1) = 91$. This tells us that $f(x) = 91$, as desired.

Finally, we have to deal with the case $x < 90$. As before, $f(x) := f(f(x + 11))$. $x + 11$ is now 100 at most, so f of it is 91, as we showed (without using the induction hypothesis!). $f(91) = 91$, as we showed in the previous paragraph. This completes the proof.

Exercise 12

$\{\langle 2n - 1, 2n \rangle : 0 < n \in \mathbb{N}\} \cup \{\langle 2n, n \rangle : n \in \mathbb{N}\}$.

Chapter 3: Partially ordered sets

Exercise 16

Set $A = \{x : f(x) \geq x\}$ and $a = \bigvee A$. ($\bot \in A$). If $x \in A$, then $x \leq a$, whence $f(x) \leq f(a)$ (f is monotone), so $x \leq f(x) \leq f(a)$ for all x in A.

So, since a is $\bigcup A$, we must have $a \le f(a)$. Since f is order-preserving, this gives us $f(a) \le f(f(a))$ (i.e., $f(f(a)) \ge f(a)$!), so $f(a) \in A$, so $f(a) \le a$. So $a = f(a)$.

Exercise (iv)

By distributivity $(a \vee b) \wedge (a \vee c) = ((a \vee b) \wedge a) \vee ((a \vee b) \wedge c)$ and the RHS is obviously $a \vee ((a \wedge c) \vee (b \wedge c))$ (contract the first bit and expand the second), which is $a \vee (a \wedge c)$ [which contracts to a] $\vee (b \wedge c)$, which is $a \vee (b \wedge c)$.

Exercise 21

Consider the function

$$F : \mathcal{P}(X) \to \mathcal{P}(X); \quad F(A) = X \setminus g``(Y \setminus f``A),$$

where $f : X \to Y$ and $g : Y \to X$ are injections.

F is monotone (with respect to $\subseteq$) because it is the composition of two monotone functions (translation under f and under g) with two antimonotone functions (complementation with respect to X and with respect to Y), and the composition of two antimonotone functions is monotone. Therefore it will have a fixed point, A, say. Then the function

$h =: \lambda x.(\text{if } x \in A \text{ then } f(x) \text{ else } g^{-1}(x))$ is a bijection $X \to Y$. It is the union of two functions with disjoint domains and disjoint ranges and so is a function, and it is total because A is a fixed point for F.

For the second part suppose $f : X \to Y$ and $g : Y \to X$ are piecewise – Σ injections. Construct h as in the formula displayed above. Any subset of (the graph of) a piecewise – Σ map is piecewise – Σ, and that the inverse of a piecewise – Σ *injection* is likewise a piecewise – Σ injection. Clearly the union of two disjoint piecewise – Σ injections is piecewise – Σ. So h is piecewise – Σ. It is total because A is a fixed point for F.

(iii) No. The open unit interval and the closed unit interval inject in order-preserving ways into each other, but they are not iso.

Exercise 24

(i) Yes. $\{n : (\exists xy \in \mathbb{N})(n = 2^x \cdot 3^y)\}$.
(ii) No. But any superset of the evens is a fixed point.

(iii) Yes. $\{n : (\exists x \in \mathbb{N} \setminus \{0\})(n = 2^x)\}$.

(iv) Yes. F is idempotent, and all terminal segments are fixed.

(v) No. (Obviously!!)

Exercise 34

If $\mathcal{B}$ is a boolean algebra with nonprincipal filters, then it has a nonprincipal ultrafilter.

The idea is to use Zorn on the chain-complete poset of nonprincipal filters. The only hard part is to show that every nonprincipal filter that is not ultra can be properly extended to another nonprincipal filter. Suppose F is a nonprincipal filter $\subseteq \mathcal{B}$, and suppose that whenever $x \notin F$, the filter generated by $F \cup \{x\}$ is principal. F is not ultra, so we can find $x \in \mathcal{B}$ with $x \notin F$ and $\neg x \notin F$. The two filters generated by $F \cup \{x\}$ and $F \cup \{\neg x\}$ are both principal, so there are $a, b \in F$ such that $a \wedge x$ and $b \wedge \neg x$ are the generators of the two principal filters. But then everything in $F \geq (a \wedge x) \vee (b \wedge \neg x)$, and this object $\geq a \cap b$, which is a member of F, so F was the principal filter generated by $a \cap b$.

Exercise 40(ii)

No, it is in fact *antimonotone*: $X \subseteq Y \to Y^\vee \subseteq X^\vee$. It is also continuous in the sense that, if $\mathcal{X}$ is a family of subsets of $\mathcal{P}(U)$, then

$$\left(\bigcup \mathcal{X}\right)^\vee = \bigcap \{X^\vee : X \in \mathcal{X}\}$$

and

$$\left(\bigcap \mathcal{X}\right)^\vee = \bigcup \{X^\vee : X \in \mathcal{X}\}.$$

If $x \in \left(\bigcup \mathcal{X}\right)^\vee$ with $y \in X \in \mathcal{X}$, then $x \cap y \neq \emptyset$, so $x \in X^\vee$, so $x \in \bigcap \{X^\vee : X \in \mathcal{X}\}$ and conversely. The second equality is proved similarly.

(iv) The poset given in the hint is chain-complete. Let $\langle C, \subseteq \rangle$ be a chain in it. We will show that $\bigcup C \subseteq (\bigcup C)^\vee$. The only interesting case is where $\bigcup C \notin C$. Fix $c \in C$. We will show that $(\forall c_j \in C)(c \subseteq c_j{}^\vee)$.

Either $c \subseteq c_j \in C$, in which case $c \subseteq c_j \subseteq c_j{}^\vee \subseteq c^\vee$, so $c \subseteq c_j{}^\vee$, or $c_j \subseteq c$, in which case $c \subseteq c^\vee \subseteq c_j{}^\vee$.

So $c \subseteq c_j{}^\vee$ for all $c_j \in C$, so $\bigcup C \subseteq c_j{}^\vee$ for all $c_j \in C$, whence $\bigcup C \subseteq (\bigcup C)^\vee$.

So by Zorn's lemma, the poset has a maximal element, X, say. By maximality of X we know that if $x \in (U \setminus X)$, then $X \cup \{x\} \not\subseteq (X \cup \{x\})^\vee$.

So there is something in $X \cup \{x\}$ that is not in $(X \cup \{x\})^\vee$ and does not meet everything in X. Since everything in X meets everything in X, this thing is obviously x. But x was arbitrary, so X already contains all things meeting all its members. So $X = X^\vee$.

(v) If $X = X^\vee$, then the intersection of any two things in X is nonempty. If there are $Y \subseteq U$ such that neither Y not $U \setminus Y$ are in X, then there is something in X disjoint from Y and something in X disjoint from $U \setminus Y$, and the intersection of these two things must be nonempty, which is impossible.

Clearly all $X^\vee$ are upper sets – closed under superset. If $|U| = 2n+1$, then $\{X \subset U : |X| \geq n+1\}$ is a fixed point.

Equally clearly, a principal ultrafilter is a fixed point.

Exercise 41

How many bases are there for the free boolean algebra of size 2^{2^n}? The first element of the basis must be above half of the 2^n atoms, and there are $\binom{2^n}{2^{n-1}}$ ways of choosing it. We have now split the (set of) atoms into two pieces. The second element must be above precisely half of the elements in each of the two pieces, and – since the choices are independent – there are $\binom{2^{n-1}}{2^{n-2}}^2$ ways of doing this. We have now split the (set of) atoms into four pieces. The third element must be above precisely half of the elements in each of the four pieces, and – since the choices are independent – there are $\binom{2^{n-2}}{2^{n-3}}^4$ ways of doing this. And so on! When multiplying them out remember that $2^n - 2^{n-1} = 2^{n-1}$, so everything except $2^n!$ disappears. Divide by $n!$ at the end because we are not interested in the order in which basis elements appear. This gives $2^n!/n!$.

Chapter 4: Propositional calculus

Exercise 43

$\bot \to ((p \to \bot) \to \bot)$ is an instance of K. $(((p \to \bot) \to \bot) \to p)$ is double negation. By composition we get $\bot \to p$, as desired. We have to justify this as follows:

$0 \ A \to B$
$1 \ B \to C$
$2 \ (B \to C) \to (A \to (B \to C))$ $\hspace{4cm}(K)$
$3 \ A \to (B \to C)$ $\hspace{5cm}(1,2, \text{MP})$

4 $(A \to (B \to C)) \to ((A \to B) \to (A \to C))$ (S)

5 $(A \to B) \to (A \to C)$ (3,4, MP)

6 $A \to C$

Exercise 44

We will treat only the hard case where T is not finitely axiomatisable. Let $\{B_i : i \in \mathbb{N}\}$ be an axiomatisation of T, and let ψ be an arbitrary theorem of T.

Now consider the new axiomatisation of T devised as follows: ψ is an axiom and $\psi \to B_i$ is an axiom for all i.

This is an axiomatisation of T, since every B_i can be proved from it. We wish ψ to be an independent axiom in this presentation. We will show that ψ cannot be proved unless it is valid. Suppose ψ is *not* independent. Then there is a finite $I \subset \mathbb{N}$ such that

$$\{\psi \to B_i : i \in I\} \vdash \psi,$$

which is to say

$$\psi \to (\bigwedge_{i \in I} B_i) \vdash \psi,$$

whence

$$(\psi \to (\bigwedge_{i \in I} B_i)) \to \psi.$$

But

$$((\psi \to (\bigwedge_{i \in I} B_i)) \to \psi) \to \psi$$

is an instance of Peirce's law, so by *modus ponens* we infer ψ. So ψ was valid.

Exercise 47

Let $\mathcal{B}$ be an arbitrary boolean algebra. Every 2-valuation is also a $\mathcal{B}$-valuation (since the two-element algebra is a subalgebra of $\mathcal{B}$), so if A is $\mathcal{B}$-valid, then it is certainly 2-valid. For the other direction suppose that v_B is a $\mathcal{B}$-valuation sending ψ to something $b \in \mathcal{B}$, where $b \neq 1$. By the prime ideal theorem there is $\mathcal{U}$ an ultrafilter in $\mathcal{B}$ with $b \notin \mathcal{U}$. Let $f_{\mathcal{U}}$ be the corresponding homomorphism onto the two-element boolean algebra. Then every $\mathcal{B}$-valuation v corresponds to a 2-valuation, namely, $f_{\mathcal{U}} \circ v$. So $f_{\mathcal{U}} \circ v_{\mathcal{B}}$ is a 2-valuation not satisfying ψ. So ψ is not 2-valid.

This shows that $\mathcal{B}$-validity is the same as 2-validity for all boolean algebras $\mathcal{B}$. For the converse, observe that any lattice that validates all 2-valid formulæ will validate $p \vee \neg p$, so it is complemented; it validates $(p \vee (q \wedge r)) \longleftrightarrow ((p \vee q) \wedge (p \vee r))$, so it is distributive, and so on.

Exercise 48

Peirce's law is not deducible from K and S.

Axiom K: $A \to (B \to A)$.

Axiom S $(A \to (B \to C)) \to ((A \to B) \to (A \to C))$.

We will need the following three-valued truth-table for the connective '$\to$'.

$\to$	1	2	3
1	1	2	3
2	1	1	3
3	1	1	1

Notice that if A and $A \to B$ both take truth-value 1, so does B. Notice also that K and S take truth-value 1 under all assignments of truth-values to the letters within them. So if ϕ is deducible from K and S, it must take value 1 under any assigment of truth-values to the literals within it (by structural induction on the family of proofs).

Then check that if A is given truth-value 2 and B is given truth-value 3, then $((A \to B) \to A) \to A$ gets truth-value 2.

So Peirce's law is not deducible from K and S.

Notice that four figures in the bottom-right corner form a copy of the ordinary two-valued table, with 3 as false and 1 as true; also, the four in the top left form a copy with 1 as true and 2 as false.

The moral of this example is that some kinds of mathematic really need formalisation. Unless we had a concept of proof, and a proof by induction on the structures of proofs, we would have no way of demonstrating that $((A \to B) \to A) \to A$ cannot be derived from K and S other than merely persistently trying and failing.

Exercise 51

(Prove the order extension principle by propositional compactness.) Suppose we have a strict partial ordering $\langle X, R \rangle$ and we seek a total order extending R. For each $a \neq b$ in X we invent a propositional letter p_{ab} and have axioms $p_{ab} \to p_{bc} \to p_{ac}$ and p_{ab} XOR p_{ba}, and we adopt p_{ab} as an axiom iff $R(a, b)$. Every finite subset of this theory is consistent. So by compactness the whole thing is consistent. A valuation gives us a strict total order of X.

Chapter 5: Predicate calculus

Exercise 53

Give sets of axioms in suitable first-order languages (to be specified) for the following theories:

(i)   The theory of integral domains;
ring axioms
$$(\forall xyz)(x \cdot (y + z) = x \cdot y + x \cdot z)$$
$$(\forall xy)(x \cdot y = y \cdot x)$$
$$(\forall x)(x \cdot 1 = x)$$
$$(\forall x)(x + 0 = x)$$
$$(\forall x)(x \cdot 0 = 0)$$
the integral domain axiom:
$$(\forall xy)(x \cdot y = 0 \to (x = 0 \lor y = 0)).$$

(ii) The theory of ordered groups (i.e., groups having a given total order): add to the group axioms the single extra axiom $(\forall xyz)(x < y \to (x \cdot z < y \cdot z) \land (z \cdot x < z \cdot y))$.

(iii) The theory of groups of order 60: write out a multiplication table:
$$(\exists x_1 \ldots x_{60})((\bigwedge_{i \neq j} x_i \neq x_j) \land (\forall y)(\bigvee_{i \leq 60} y = x_i) \land \ldots)$$

(iv) The theory of simple groups of order 60. There is only one simple group of order 60, namely, A_5. Write out a multlication table for it. Do not try adding to the axioms of group theory an axiom asserting that there are precisely 60 objects and axioms to say the group is simple because there is no first-order set of axioms that says that a group is free. This can be shown using ultraproducts. Take an ultraproduct of all the cyclic groups of order p for p prime. If the ultrafilter is nonprincipal, the ultraproduct is not simple, since it is an infinite abelian group and all its subgroups are normal.

(v) The theory of algebraically closed fields of characteristic 0. Add infinitely many axioms – one for each prime p – to say that the field is not of characteristic p.

(vi) The theory of partial orders in which every element belongs to a unique maximal antichain is axiomatised by adding to the theory of posets the condition in exercise 33. The significance of this is that "every element belongs to a unique maximal antichain" is on the face of it a second-order property. But it turns out to be first-order after all!

(vii) The theory of commutative local rings (a local ring being a ring with a unique maximal ideal). If there is a unique maximal ideal, it turns out that it must be the ideal of noninvertible elements. So all you need to do is say that the noninvertible elements form an ideal.

Exercise 54

Which of the following have first-order theories?

(i) Groups all of whose elements are of finite order? No: compactness.

(ii) Groups all of whose nonidentity elements are of infinite order? Yes: for each concrete k have an axiom saying that every nonzero element is of order k at least.

(iii) Groups with trivial centre? Yes: $(\forall x)(x \neq 1 \rightarrow (\exists y)(xy \neq yx))$.

(iv) Groups with an element of infinite order in their centre? Yes: add to the language of group theory a single constant a, an axiom to say that a is in the centre and infinitely many axioms to ensure that it is of infinite order.

(vi) Noetherian rings (rings wherein every $\subseteq$-chain of ideals has a maximal element). This is obviously higher-order, but the proof is not straightforward. The definition of Noetherian makes it sound third-order, but this is equivalent to a condition that sounds more second-order: every ideal is finitely generated. Is this really second-order?

(vii) The theory of free groups is not first-order, but a proof is beyond the scope of this book.

(viii) Torsion-free abelian groups. Infinitely many first-order axioms will capture torsion-freeness.

Exercise 55

Given $X \subseteq \mathbb{N}$, devise T_X as follows. For $n \notin X$ put into T_X the axiom that says there are not exactly n objects. Clearly any finite model of T_X cannot have size n for $n \notin X$.

Exercise 74

A *pedigree* is a set P with two unary total functions f amd m defined on it, with disjoint ranges. ($m(x)$ is x's mother and $f(x)$ is x's father.)

(i) Set up a first-order language $\mathcal{L}$ for pedigrees and provide axioms for a theory T_1 of pedigrees.

The obvious answer is two unary function letters f and g. Axioms will be $(\forall x)(\forall y)(f(x) \neq m(y))$. One could use two binary predicate letters instead, but then one needs axioms like $(\forall x)(\exists y)(M(x, y))$ and suchlike.

A pedigree may be **circle-free**: in a realistic pedigree no one is their own ancestor! Realistic pedigrees are also **locally finite**: no one is the father or mother of infinitely many things.

(ii) One of these two new properties is first-order and the other is not. Give axioms for a theory T_2 of the one that is first-order and an explanation of why the other one is not.

There is an infinitely axiomatisable theory of circle-free pedigrees. It has a lot of axioms like $(\forall x)(x \neq f(x))$, $(\forall x)(x \neq m(x))$, $(\forall x)(x \neq m(f(x)))$, and so on. There is a simple compactness argument to show that the theory of locally finite pedigrees is not first-order.

A *fitness function* is a map v from P to the reals satisfying $v(x) = (1/2) \cdot \Sigma_{f(y)=x} v(y)$ or $v(x) = (1/2) \cdot \Sigma_{m(y)=x} v(y)$ (depending on whether x is a mother or a father).

(iii) Find a sufficient condition for a pedigree to have a nontrivial fitness function, and a sufficient condition for it to have no nontrivial fitness funtion.

For any pedigree $\langle P, f, m \rangle$ whatever, the set $P \to \Re$ forms a vector space. The identically zero function is of course a fitness function, and if v is a point in this space, then v' defined by $\lambda x.(1/2) \cdot \Sigma_{f(y)=x} v(y)$ or $v(x) = (1/2) \cdot \Sigma_{m(y)=x} v(y)$ (depending on whether x is a mother or a father) is another, and clearly something is a fitness function iff it is a fixed point for this operation. Obvious though this approach is, it appears to me at least to be less use than one might expect.

If the limit of the number of lineages of length n starting at x divided by 2^n is well-defined, take this to be $v(x)$. Notice circle-free is not important here.

(iv) Extend your language $\mathcal{L}$ to include syntax for v. In your new language provide axioms for a new theory T_3 that is to be a conservative extension of T_1 and whose locally finite models are precisely the locally finite pedigrees with a nontrivial fitness function.

For each concrete n add an axiom saying, "For all x, if x has precisely n children, then $v(x)$ is half the sum of the v's of those children". This can be done in a first-order way.

There is an obvious concept of *generation* for a pedigree.

(v) Expand $\mathcal{L}$ by adding new predicate(s), and give a first-order theory in this new language for pedigrees that have well-defined generations. Give first-order axioms in $\mathcal{L}$ itself for a theory of pedigrees that have well-defined generations.

We can capture the presence of generations in a first-order way with infinitely many axioms like $(\forall xyz)(\neg(x = f(y) \wedge x = f(z) \wedge z = m(y)))$. One can also use ideas from graph theory, specifically ideas about how to axiomatise colouring of digraphs. Invent a new one-place predicate letter G, where the intended meaning of $G(x)$ will be that x belongs to a generation that is an even distance from some arbitrary origin. Then one needs only the axiom $(\forall x)(G(x) \longleftrightarrow (\neg G(f(x)) \wedge \neg G(m(x))))$. This makes the point that enlarging the language often enables one to give simpler axiomatisations. All this does is say that the underlying digraph is two-colourable and is compatible with everyone being their own grandparent. This is of course ruled out by the cycle-free axioms. But we can also rule it out by having more colours G_i, with a family of axioms indexed by the subscripts saying $(\forall x)(G_{i+1}(x) \longleftrightarrow (G_i(f(x)) \wedge G_i(m(x))))$ with corrections for overflow or underflow if the set of generations is to be finite. If it is infinite, then of course one cannot say in a first-order way that everything belongs to a generation! However, if one opts for a two-sorted theory with quantifiers over subscripts, one can say things like $(\forall x)(\exists i)(G_i(x))$. This ensures that everything belongs to a generation, but an ultrapower will of course give us a model in which the generations are indexed by nonstandard integers.

Chapter 6: Computable functions

Exercise 78

78.(i) Find a primitive recursive declaration for the function commonly declared by

$$f(0) := f(1) := 1; \quad f(n+1) := f(n) + f(n-1).$$

Declare $F(0) := \langle 1, 1 \rangle; \quad F(n+1) = \langle \mathtt{snd}(F(n)), \mathtt{fst}(F(n)) + \mathtt{snd}(F(n)) \rangle$

Then $\mathtt{Fib}(n) = \mathtt{fst}(F(n))$.

This technique is commonly called **pipelining**.

78.(ii) We want to represent H as something obtained by iteration. The function we are going to define by iteration will be $\lambda n.\langle H(n), n \rangle$ (though of course that is not how it is *explicitly* defined!), and then we get H from it by composition ($\mathtt{fst}$). Abbreviate $\lambda n.\langle H(n), n \rangle$ to F. Then we have

$$\begin{aligned} F(S(y)) &= \langle H(S(y)), S(y) \rangle \\ &= \langle G(H(y), y), S(y) \rangle \end{aligned}$$

(we know this by the recursion). Now $H(y) = \mathtt{fst}(F(y))$ and $y = \mathtt{snd}(F(y))$, so this is

$$\langle G(\mathtt{fst}(F(y)), \mathtt{snd}(F(y))), S(\mathtt{snd}(F(y))) \rangle,$$

and we notice that all occurrences of 'y' are wrapped up in F's, so this is

$$f(F(y)),$$

where f is

$$\lambda z.\langle G(\mathtt{fst}z, \mathtt{snd}z), S(\mathtt{snd}z) \rangle,$$

so $H(y) = \mathtt{fst}(Fy) = \mathtt{fst}(f^y(F0)) = \mathtt{fst}(f^y(b))$, where $b = \langle a, 0 \rangle$.

Exercise 85, part (i)

If T has a semidecidable set of axioms, then the set of its theorems is also semidecidable. This is because the relation of deducibility between (gnumbers of) formulæ is primitive recursive, so the set of theorems of T is the extension of a predicate $T(x)$ that informally says $(\exists y)(\exists z)(y$ is a run of a Turing machine that emits a list of (some) axioms of T and z is a derivation of x from those axioms). Since the matrix of this formula is primitive recursive, and we can always code two naturals as one, this is of the form $(\exists w)(\phi)$, where ϕ is primitive recursive and the extensions of all such predicates are semidecidable sets. So the set of theorems of T is a semidecidable set, just like the set of theorems of a theory with a *recursive* axiomatisation! This indeed is the burden of question 6.6.98. That is why we would *expect* the truth of the thing we have been invited to prove, but sadly that is not how we prove it.

T has a semidecidable axiomatisation, $\{S_i : i \in \mathbb{N}\}$, say. Without loss of generality I can assume that none of these axioms are anything silly like $p \wedge p$ or $p \wedge p \wedge p$, by simply deleting extra copies. (Strange to say, this will matter!) That is to say, there is a Turing machine $\mathcal{M}$ that emits members of $\{S_i : i \in \mathbb{N}\}$ in that order when run in parallel with itself in the style of remark 43. Now define a new set $\{S_i' : i \in \mathbb{N}\}$ of axioms by setting S_i' to be the conjunction of i copies of S_i. Clearly this is a semidecidable set, for it is simple enough to tweak $\mathcal{M}$ into something whose ith announcement is S_i' and not S_i. We have to show that its complement is semidecidable as well, that is, that we can recognise formulæ that are not in $\{S_i : i \in \mathbb{N}\}$. Here is how I do it. You give me a formula Ψ. I count how many conjunctions it has at the top level: i, say, so that it is $\psi \wedge \psi \wedge \ldots \psi$, i times. I run $\mathcal{M}$ until it has emitted i formulæ, and I check to see whether or not the ith formula is ψ. If it is, Ψ is an axiom, and if it is not, it is not!

Exercise 85, part (ii)

See Bunder (1989)

Exercise 87 (1991:5:10 (CS))

Define the terms *primitive recursive function; partial recursive function; total computable function.*

Ackermann's function is defined as follows:

$$A(0,y) := y+1; \; A(x+1,0) := A(x,1); \; A(x+1,y+1) := A(x, A(x+1,y)).$$

For each n define $f_n(y) := A(n,y)$.

Show that, for all $n \geq 0$, $f_{n+1}(y) = f_n^{y+1}(1)$, and deduce that each f_n is primitive recursive. Why does this mean that the Ackermann function is total computable?

You have to prove by induction on 'y' that this holds for all n.

Base case: $y = 0$. We want $(\forall n)(f_{n+1}(0) = f_n^1(1))$. Let n be arbitrary. Want $f_{n+1}(0) = f_n^1(1)$. Expand using definition of f:

LHS: $f_{n+1}(0) = A(n+1,0) = A(n,1)$. RHS: $f_n^1(1) = A(n,1)$, as desired.

Now for the induction step.

Assume $(\forall n)(f_{n+1}(y) = f_n^{y+1}(1))$. We want $(\forall n)(f_{n+1}(y+1) = f_n^{y+2}(1))$. Let n be arbitrary as before and expand $f_{n+1}(y+1)$ as before to get $A(n+1, y+1)$, which is $A(n, A(n+1, y))$, which is $f_n(f_{n+1}(y))$. But by induction hypothesis on 'y', $f_{n+1}(y) = f_n^{y+1}(1)$, so $f_n(f_{n+1}(y)) = f_n(f_n^{y+1}(1))$, which of course is $f_n^{y+2}(1)$.

This is a useful example of a general problem. There are two universally quantified variables in '$(\forall n)(\forall y)(n \geq 0 \rightarrow f_{n+1}(y) = f_n^{y+1}(1)$, and both of them range over a rectype. On the face of it, each quantifier can be dealt with by either a UG or an induction. In many case, like the one in hand, there is only one strategy that will work. You *have* to treat 'n' by UG and 'y' by induction.

Finally, we show that f_n is primitive recursive for each n. We will use the fact we have proved, namely, that $f_{n+1}(y) = f_n^{y+1}(1)$. Consider the declaration:

$$g(0) := h(1); \quad g(n+1) = h(g(n)).$$

This is clearly primitive recursive: g will be primitive recursive if h is. But f_{n+1} is obtained by primitive recursion over f_n in precisely the way g is declared over h, so, as long as f_0 is primitive recursive, we can prove all the f_n to be primitive recursive by induction on n.

Exercises 88 and 90

These two questions make the same point and should be considered together.

Exercise 88

The second function is a step function and is therfore computable: all step functions are. The fact that we do not know *which* step function it is merely means that we do not know how to compute it.

The first function might be computable. Nobody has a clue. My guess is that it is the computable function $\lambda n.\mathtt{true}$.

The third function is in fact computable, in the sense that there is a computable function with the same graph, but you would not guess it from the declaration. Remember that computability is in the first instance a property of function declarations (functions-in-intension), not of functions-in-extension.

Exercise 90

Either the function f is eventually constant, in which case its range is finite and is therefore computable, or it is unbounded. If it is unbounded, the way to test whether or not the candidate number is a value of f is to compute f of 0,1,2,3 ... until the candidate number is either hit or overtaken. Of course, if you do not know which of these situations is the one you are in, you have no way of discovering the decision method in virtue of which this set is decidable, but that is your problem, not God's. This rams home the point that, for a problem to be solvable, what is necessary is that there should be a decision method for it – not for there to be a decision method for it *known to us*.

You might think this is an elementary point, and it is, but it is one that can be easily overlooked.

Exercise 100

Let $\{\phi_n | n \in \mathbb{N}\}$ be an enumeration of the partial computable functions of arity 1, that is, $\phi(n) : \mathbb{N} \to \mathbb{N}$.

Now take $f : \mathbb{N} \to \mathbb{N}$ to be $f(m) = \phi_m(m) + 1$. Note that f is certainly not total, and it certainly is recursive. Suppose now that h is total computable and extends f to all of $\mathbb{N}$. Then we must have $h = \phi_{n_0}$ for some n_0, because h is total computable.

Then $h(n_0) = \phi_{n_0}(n_0)$, and as h total, the latter is defined. Therefore $f(n_0)$ is defined, and

$$\phi_{n_0}(n_0) = h(n_0) = f(n_0) = \phi_{n_0}(n_0) + 1.$$

Exercise 104

Let $[X]^n$ be the set of unordered n-tuples of members of X. Consider $\rho: [\mathbb{N}]^3 \to \{0,1\}$ by $\rho(\{x,y,z\}) = 0$ iff $x < y < z \to (\forall p, d < x)[\{p\}_y(d) \downarrow \longleftrightarrow \{p\}_z(d) \downarrow]$ (where $\{p\}_z(d) \downarrow$ means that the program with gnumber p applied to data d halts in $\leq z$ steps).

Suppose X is an infinite subset of $\mathbb{N}$ homogeneous for ρ. We must have $\rho``[X]^3 = \{0\}$ since $\rho``[X]^3 = \{1\}$ is obviously impossible ("too few truth-values"). We will show that, if X is recursive, then we can solve the halting problem.

To determine whether or not $\{p\}(d) \downarrow$, first find a member n of X larger than p and d. Then for any y, z in X bigger than n we have $\{p\}_y(d) \downarrow$ iff $\{p\}_z(d) \downarrow$. Since X is infinite, it has arbitrarily large members and so if $\{p\}(d)$ ever halts at all, there is $z \in X$ large enough to ensure that $\{p\}_z(d) \downarrow$. But then, by homogeneity of X, it will be sufficient to check $\{p\}_z(d) \downarrow$ for even *one* $z \in X$ bigger than n.

On the other hand, there is a theorem of Seetapun's (see Hummel 1994) that says that every recursive partition of $[\mathbb{N}]^2$ has a homogeneous set in which the halting problem is not recursive.

Chapter 7: Ordinals

Exercise 105

Let us say $\beta \in On$ is **normal** if $(\forall \gamma \in On)(\gamma <_{On} \beta \to \text{succ } \gamma \leq_{On} \beta)$. (We will use this terminology only in the next three propositions.)

PROPOSITION 107 *If β is normal, then* $(\forall \gamma \in On)(\gamma \leq_{On} \beta \lor \text{succ } \beta \leq_{On} \gamma)$.

Proof: That is to say, we show that – for all normal β – $\{\gamma \in On : \gamma \leq_{On} \beta \lor \text{succ } \beta \leq_{On} \gamma\}$ contains 0 and is closed under succ and sups of chains and is therefore a superset of On. Let us deal with these in turn.

   (i) (Contains 0). $0 \in \{\gamma \in On : \gamma \leq_{On} \beta \lor \text{succ } \beta \leq_{On} \gamma\}$ because $0 \leq_{On} \beta$ by clause 4 of definition 51.

   (ii) (Closed under succ). If $\gamma \in \{\gamma \in On : \gamma \leq_{On} \beta \lor \text{succ } \beta \leq_{On} \gamma\}$ then either

   (a) $\gamma <_{On} \beta$, in which case $\text{succ } \gamma \leq_{On} \beta$ by normality of β and $\text{succ } \gamma \in \{\gamma \in On : \gamma \leq_{On} \beta \lor \text{succ } \beta \leq_{On} \gamma\}$; or

   (b) $\gamma = \beta$, in which case $\text{succ } \gamma \leq_{On} \text{succ } \beta$ and $\gamma \in \{\gamma \in On : \gamma \leq_{On} \beta \lor \text{succ } \beta \leq_{On} \gamma\}$; or

   (c) $\text{succ } \beta \leq_{On} \gamma$, in which case $\text{succ } \beta \leq_{On} \text{succ } \gamma$ and $\text{succ } \gamma \in \{\gamma \in On : \gamma \leq_{On} \beta \lor \text{succ } \beta \leq_{On} \gamma\}$.

   (iii) (Closed under sups of chains). Let $S \subseteq \{\gamma \in On : \gamma \leq_{On} \beta \lor \text{succ } \beta \leq_{On} \gamma\}$ be a chain. If $(\forall \gamma \in S)(\gamma \leq_{On} \beta)$, then $\sup S \leq_{On} \beta$. On the other hand, if there is $\gamma \in S$ s.t. $\gamma \not\leq_{On} \beta$, we have $\text{succ } \beta \leq_{On} \gamma$ (by normality of β), so $\sup S \geq_{On} \text{succ } \beta$ and $\sup S \in \{\gamma \in On : \gamma \leq_{On} \beta \lor \text{succ } \beta \leq_{On} \gamma\}$.

PROPOSITION 108 $\leq_{On}$ *is total.*

Proof: It will suffice to prove that every ordinal is normal. Naturally we do this by induction: the collection of normal ordinals will contain 0 and be closed under **succ** and **sups** of chains.

(i) (Contains 0). Vacuously!

(ii) (Closed under **succ**). Suppose $\beta \in \{\zeta \in On : (\forall \gamma \in On)(\gamma <_{On} \zeta \to$ **succ** $\gamma \leq_{On} \zeta\}$. We will show $(\forall \gamma \in On)(\gamma <_{On}$ **succ** $\beta \to$ **succ** $\gamma \leq_{On}$ **succ** β). So assume $\gamma <_{On}$ **succ** β. This gives $\gamma \leq_{On} \beta$ by normality of β. If $\gamma = \beta$, we certainly have **succ** $\gamma \leq_{On}$ **succ** β as desired, and if $\gamma <_{On} \beta$, we have **succ** $\gamma \leq_{On} \beta \leq_{On}$ **succ** β.

(iii) (Closed under **sups** of chains). Suppose $S \subseteq \{\zeta \in On : (\forall \gamma \in On)(\gamma <_{On} \zeta \to$ **succ** $\gamma \leq_{On} \zeta)\}$ is a chain. If $\gamma <_{On}$ **sup** S, we cannot have $(\forall \zeta \in S)(\gamma \geq$ **succ** $\zeta)$ for otherwise $(\forall \zeta \in S)(\gamma \geq \zeta)$ (by transitivity and inflationarity of **succ**), so for at least one $\zeta \in S$ we have $\gamma \leq_{On} \zeta$. If $\gamma <_{On} \zeta$, we have **succ** $\gamma \leq_{On} \zeta \leq_{On}$ **sup** S since ζ is normal. If $\gamma = \zeta$, then ζ is not the greatest element of S, so in S there is $\zeta' > \zeta$ and then **succ** $\gamma \leq_{On} \zeta' \leq_{On}$ **sup** S by normality of ζ'.

If β and γ are two things in On, we have $\gamma \leq_{On} \beta \vee$ **succ** $\beta \leq_{On} \gamma$ by normality of β, so the second disjunct implies $\beta \leq_{On} \gamma$, whence $\gamma \leq_{On} \beta \vee \beta \leq_{On} \gamma$. ∎

PROPOSITION 109 $<_{On}$ *is well-founded.*

Proof: The obvious thing to do is to show that the collection of nice ordinals (α is nice iff every set of ordinals containing α has a least element) is closed under **succ** and **sup**, for then every ordinal will be nice.

(i) RTP: α nice implies **succ** α nice.

Suppose **succ** $\alpha \in X$. Does X have a $<_{On}$-minimal element? It does if it contains α. Suppose it does not have a minimal element. Then $X \cup \{\alpha\}$ contains a minimal element but X does not. Clearly α is the minimal element of $X \cup \{\alpha\}$. In these circumstances we want to prove that **succ** α is the least element of X. But this is assured because, by proposition 108, every ordinal is normal, so there is nothing strictly between α and **succ** α.

(ii) RTP: If β is a **sup** of a chain C of nice ordinals, then β is nice.

Let X contain β. We seek a minimal member for X. If X

meets C, then it has a minimal element; if not, we say that either all members of X are $\geq \beta$ (in which case β is the desired minimal element) or some member γ of X is below β (by totality). But γ must now be below some member δ of C, for if it were above all of them (use totality again), it would be $\geq \beta$. But now $X \cup \{\delta\}$ has a minimal element. This minimal element cannot be δ because $\delta > \gamma \in X$. So the minimal element was in X.   ∎

Exercise 106

For each α we prove by induction on γ that $\alpha \leq_{On} \alpha + \gamma$.

For the other direction we want $\alpha \leq_{On} \beta \rightarrow (\exists \gamma)(\alpha + \gamma = \beta)$. Fix α and consider the least $\beta > \alpha$ for which there is no such γ. β is not a successor, so it is a **sup** of a set $\{\alpha + \gamma : \gamma \in X\}$, so by clause 3 of definition 52 (+ at limits), β is $\alpha+$ **sup** (X).   ∎

Exercise 108

Give a recursive definition of ordinal subtraction, and prove that your definition obeys: $\beta + (\alpha - \beta) = \alpha$:

The definition is

$0 - \alpha := 0$;

$(\text{succ } \beta) - \alpha := \text{succ}(\beta - \alpha)$;

$(\text{sup } X) - \beta := \text{sup}\{\alpha - \beta : \alpha \in X\}$.

To prove $\beta + (\alpha - \beta) = \alpha$, fix β and do it by induction on α. This is easy for $\alpha = 0$. Then $\beta + ((\text{succ } \alpha) - \beta)$

$= \beta + \text{succ}(\alpha - \beta)$

$= \text{succ}(\beta + (\alpha - \beta))$

$= \text{succ } \alpha$ by induction hypothesis

and $\beta + (\text{sup } X - \beta)$

$= \beta + \text{sup}\{\alpha - \beta : \alpha \in X\}$

$= \text{sup}\{\beta + (\alpha - \beta) : \alpha \in X\}$

$= \text{sup } X$.

Chapter 8: Set theory

Exercise 118

Show that the transitive closure $^*R$ of R is well-founded iff R is.

Proof: One direction is easy, for any R-predecessor of x in X is also a *R-predecessor of x in X, so if R is not well-founded, neither is *R.

For the other direction, the hard part is to ascertain exactly what the induction hypothesis should be. The answer is that one should prove by R-induction on 'x' that every $X \subseteq dom(R)$ containing an R-ancestor of x has an *R-minimal element.

Exercise 125

Set $\langle x, y \rangle$ to be $f\text{``}x \cup g\text{``}y$. To decode an ordered pair p observe that nothing is a value both of f and of g, so p is the disjoint union of $p \cap f\text{``}V$ and $p \cap g\text{``}V$. So its two components are $f^{-1}\text{``}(p \cap f\text{``}V)$ and $g^{-1}\text{``}(p \cap g\text{``}V)$.

This idea is Quine's. See Quine (1995 chapter VIII).

Exercise 126

(i) VN → Choice. The Burali-Forti paradox tells us that the collection of all (von Neumann) ordinals is a proper class, and that it is well-ordered (by $\in$, as it happens). Von Neumann's axiom therefore implies that the universe is in bijection with On and is well-ordered. This is the axiom of *global choice*. The axiom of (local) choice asserts that every set can be well-ordered. These two axioms of choice are equivalent under realistic assumptions.

(ii) VN → Replacement. It will be sufficient to show that no set is the same size as V, for then, if we are given that a class C is the same size as a set, it will follow that it is not the same size as V and therefore, by VN, is a set.

So we must show that no set is the same size as V. Suppose X is a set and $\pi : X \longleftrightarrow V$. Let $A = \{y \in X : y \notin \pi(y)\}$. A is a set by comprehension, and we then obtain a Russell-style paradox by asking whether or not $\pi^{-1}(A) \in A$. (We can do it more directly by saying that x must be smaller than its power set, which is no bigger than V but that needs the axiom of power set.)

(iii) Choice and replacement → VN. Now suppose the axiom of (global) choice holds. Consider the pre-well-order (see the answer to question 3.3) $\leq$ of V defined by $x \leq y$ iff $\rho(x) \leq \rho(y)$. Refine this to a well-ordering of V by well-ordering each V_α. Call this $\langle V, \leq \rangle$. Because each level is the power set of the union of all preceding levels, Cantor's theorem assures us that the order type of this well-ordering is an initial ordinal, and by replacement it must be a regular ordinal as well. Now let $C \subset V$ be some arbitrary subclass of V, and think about how it manifests itself as a substructure of $\langle V, \leq \rangle$. Either (i) it is cofinal in V, in which case, by regularity, it must be the same size as V; or (ii) it is not. If it is not, the rank of its members is bounded by some ordinal α, and therefore $C \subseteq V_\alpha$ and is a set by comprehension. ∎

Bibliography

Aczel P., (1988) Non-Well-Founded Sets, CSLI Lecture Notes, *14* Stanford, California, (1988)

Ahlfors, L. V. (1953) Complex Analysis. McGraw-Hill, Toronto and London 1953

Aigner, M. and Ziegler, (2001) G. M. Proofs from The Book, 2nd edn, Springer 2001

Berlekamp, E. Conway, J.H. and Guy, R.K. (1982) Winning Ways (2 vols) Academic press

Bunder, M. Journal for Non-Classical Logic *6* 1989 pp 57–62.

Cajori, F. (1993) A history of Mathematical Notations Dover publ New York, 1993

Conway, J. H. (2001) On Numbers and Games, 2nd edn. A.K.Peters

Davenport, H. (1999) The Higher Arithmetic, 7th edn, Cambridge University Press.

Easton, W. B. (1970) Powers of regular cardinals, Annals of Mathematical Logic **1** 139 – 178.

Forti, M. and Honsell, F. [1983] Set theory with free construction Principles. *Annali della Scuola Normale Superiore di Pisa, Scienze fisiche e matematiche* **10**: 493–522.

Gardner, M. (1961) More mathematical puzzles and diversions. Penguin Books, London.

Garey, M. R. and Johnson, D.S. (1979) Computers and Intractability: a guide to the theory of NP completeness. W. H. Freeman

Girard, J-Y., Lafont, Y. and Taylor, P. T. (1989) Proofs and Types. Cambridge Tracts in Theoretical Computer Science *7* Cambridge University Press

Hardy, G. H. (1949) Divergent Series. Oxford, The Clarendon Press

Henson, C. W. (1973) Permutation methods applied to NF. *Journal of Symbolic Logic* **38** pp. 69–76.

Hofstader, D. (1979) Gödel, Escher, Bach. Basic Books

Hume, D. (1739) A Treatise of Human Nature.

Hummel, T. L. (1994) Effective versions of Ramsey's theorem: avoiding the cone above $0'$. *Journal of Symbolic Logic 59* pp 1301 – 1325.

Johnstone, P.T. (1982) Stone Spaces. Cambridge University Press

Johnstone, P.T. (1987) Notes on Set Theory. Cambridge University Press

Keisler, H. Jerome. (1976) Elementary calculus: Prindle, Weber & Schmidt, Boston (and at http://www.math.wisc.edu/ keisler/calc.html)

Keisler, H. Jerome (1976b) Foundations of infinitesimal calculus: Prindle, Weber & Schmidt, Boston

Diogenes Laertius, *Lives of the Philosophers* Book 6, Chapter 97.

Leśniewski, S. (1929) Grundzüge eines neuen Systems der Grundlagen der Mathematik, *Fundamenta Mathematicæ 14* 1 – 81, English translation (Fundamentals of a new system of the foundations of mathematics) in Stanisław J. Surma, Jan T. Srzednicki, D. I. Barnett and V. Frederick Rickey (eds) (1992) Stanisław Leśniewski: Collected Works, Kluwer.

Martin D. A. (1975) A new proof of Borel Determinacy. *Annals of Mathematics 102*: 365-71.

Mendelson, E. (1979) Introduction to Mathematical Logic. (2nd edn) Van Nostrand

Popper, K. (1968) The Logic of Scientific Discovery. Hutchinson, London

Prior, A. (1976) Papers in Logic and Ethics (P.T.Geach and A.J.P Kenney, eds). Duckworth, London.

Quine, W. V. (1962) Mathematical Logic (revised edition) Harper Torchbooks 1962

Quine, W. V. (1995) Selected Logic Papers. Harvard University Press 1995.

Russell, B. A. W. (1919) Introduction to Mathematical Philosophy.

Russell, B. A. W. and Whitehead, A. N. (1913) Principia Mathematica. Cambridge University Press

Ryle, G. (1963) The concept of mind Penguin Books, London

Silver, J. [1974] "On the singular Cardinals problem," Proceedings of the International Congress of Mathematicians pp 265 – 268.

Smullyan, R. (2001) Equivalence Relations and Groups, in *Logic, Meaning and Computation: essays in memory of Alonzo Church*, C.A. Anderson and M. Zelëny eds, Synthese library **305**. Kluwer, Dordrecht, Boston and London 2001 pp 261-274

Tarski, A. (1986) *What are Logical Notions? History and Philosophy of Logic 7* (1986) pp 143 – 154.

Wang, H. (1949) On Zermelo's and von Neumann's axioms for Set Theory. *Proc. Nat. Acad. Sci. 35*: pp 150–155.

Index

233